## Introduction

## A: Legal and management



## B: Health and welfare



## C: General safety



## D: High risk activities

## E: Environment



## F: Specialist activities



## Further information

# Introduction

# Overview

Good communication is essential for the management of health and safety on building and construction sites. It is vitally important that contractors, managers and supervisors engage and consult with the workforce, as it is an effective way of identifying hazards and controlling the risks.

It is the workforce who have first-hand experience of actually carrying out the job and who may therefore be more knowledgeable about the risks involved.

Willing and effective worker participation and feedback will be achieved when the workforce:

- ☑ is consulted with
- ☑ has confidence in their managers and supervisors
- ☑ knows their ideas and concerns will be listened to and, if required, acted upon
- ☑ has sufficient knowledge to recognise when something is not safe
- ☑ has been trained in the skills necessary to deliver effective feedback.

Ample opportunity must be given for all members of the workforce to consult with management on any concerns that they have regarding health and safety.

Toolbox talks are one way of delivering advice on matters of health, safety and the environment, as well as engaging in discussions to obtain feedback, thus helping to maintain and improve standards. However, they are not a substitute for formal training in the correct methods of instruction.

*Toolbox talks* (GT 700) has been designed to assist supervisors at all levels to prepare and deliver effective toolbox talks on building and construction sites.

# Acknowledgements

CITB wishes to acknowledge the assistance offered by the following organisations in the preparation of this edition of GT 700:

- ☑ Carillion
- ☑ Costain
- ☑ eBrit Services
- ☑ Environment Agency
- ☑ J Breheny Contractors Ltd
- ☑ MJ Fuller & Associates
- ☑ Morgan Sindall
- ☑ Skanska.

# About the construction industry

These facts can help to support your toolbox talks.

- ☑ Approximately 2.47 million people are employed in the UK construction industry. This includes housing, utilities, repair and maintenance, refurbishment, shopfitting, demolition, roofing, mechanical and electrical, plumbing and highways maintenance.
- ☑ The UK construction industry is made up of 197,965 construction firms and 90% of companies employ less than 10 workers.
- ☑ Construction workers (just like you) will die due to work-related ill health. Work-related respiratory disease covers a range of illnesses that are caused or made worse by breathing in hazardous substances, such as construction dust, that damages the lungs.
- ☑ 4,500 people die each year from asbestos-related lung diseases.
- ☑ 500 people (and more each year) are dying from silica-related lung diseases (for example, dust from cutting blocks and kerbs). Many more suffer from occupational asthma or are forced to leave the industry due to work-related ill health.

- ☑ On average 50 workers are killed each year due to accidents. The biggest killer (around half) is due to falls from height. Around 7 fall through roofs and roof lights.
- ☑ Each year approximately 2,500 people are seriously injured (for example, suffering broken bones, fractured skull and amputations) and 5,700 people have reportable injuries. One common cause is being struck by mobile plant, vehicles and equipment.
- ☑ The most common causes of injuries are due to manual handling and slips, trips and falls. 60% of all work at height injuries are due to falls from below head height.

## How to use Toolbox talks

GT 700 follows the standard structure that is used across all core CITB publications.

**Section A** **Legal and management**

**Section B** **Health and welfare**

**Section C** **General safety**

**Section D** **High risk activities**

**Section E** **Environment**

**Section F** **Specialist activities**

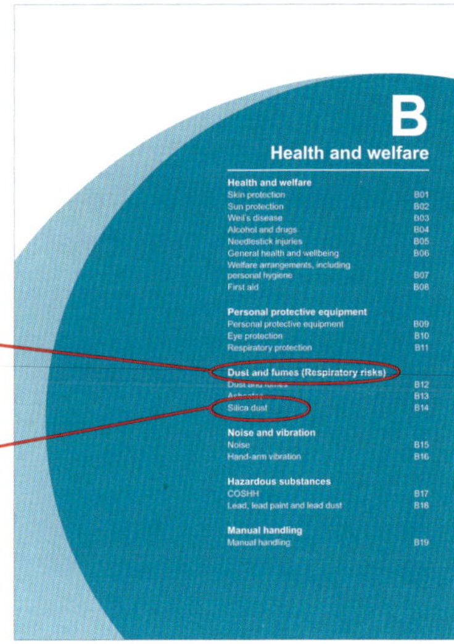


At the start of each section there is a contents list, which is divided into topic areas. Each topic area relates to a chapter within the CITB's supporting publication *Site supervision simplified* (GE 706).

The letter of each toolbox talk relates to the section that it corresponds to. For example, toolbox talk B14 is Silica dust, and it is relevant to Section B Health and welfare.

The simple structure of GT 700 will assist supervisors who have not received any formal training on how to deliver toolbox talks.

Each toolbox talk is set out over two facing pages.

The first page is aimed at you, the supervisor. It starts by outlining the reasons for the talk, followed by information that has been broken down into manageable sections. The following two icons appear at the bottom of this page.

**?** This icon lists some suggested questions that you can ask. It is important that the participants' understanding is tested by asking questions, but you should avoid asking questions that could have a 'yes' or 'no' answer – instead, ask questions that start with 'what, when, why, how, name' and so on.

**!** This icon is a prompt to encourage discussions, by using a real-life situation or example, and to ask if there are any other questions or if anyone has a similar experience to tell.

The second page generally contains images, which can help to enhance the talk, or provide you with a space to make notes. If you have a hand-out that you want to use as an example or something specific to mention, it is a good idea to write it down – you can easily forget things if you do not have a prompt.

At the end of GT 700 is a briefing record which can be copied, along with a log to record the toolbox talks you have delivered and a feedback form. These are also available on, and can be printed from, the CD-ROM. It is important that any feedback is acted upon, either to improve the working practice or to note how the toolbox talk can be improved when delivered in the future.

# Preparing and delivering a toolbox talk

It is important that you prepare in advance for any toolbox talk.

Think about if you were in the audience and you were told to attend a toolbox talk. If someone just read from a book, word for word, would you want to listen or even be there?

So, what information would you need? What would hold your attention? What questions would you ask?

The information contained within GT 700 does not have to be followed to the letter – the key points are there as a guide. You can add to them using your own experiences or leave points out if they are not relevant to your particular site or activity.

Also think about the location where you are giving the talk. Would it be best suited to being:

- ✓ in a quiet area with no distractions
- ✓ on site at the work location
- ✓ in an area suitable for using any equipment that will be demonstrated?

*Ensure your toolbox talk is engaging*

Using aids (such as samples of equipment, hand-outs, photographs) or even giving a demonstration will hold people's attention. For example, if you are giving a talk on the safe use of podium steps, then try to give the talk on site with a podium step close by (not in the canteen, with no equipment).

The suggested length of a talk is between 10 and 30 minutes. However, the duration will also depend upon the amount of interaction there is with the people receiving the talk. Unless you have to, do not stifle discussions purely because of time issues; people are obviously taking an interest and learning from it.

## Prepare – What should you consider?

1. Before the talk, think about what preparation is required.
2. Check the safety aspects of the proposed location.
3. Consider effective aids that you can use to enhance the talk.
4. During the talk, keep distractions to a minimum.

## Reason – Why are you giving the talk?

1 Employees are generally information-proof, they filter out what they can ignore. In giving a talk, you have to make sure that you gain their interest from the start.

2. Get their attention as quickly as possible. You won't get the attention of your people by saying, 'Well, it's time for another safety talk…'
3. Look for interesting opening lines. Each talk will offer a suggestion but, of course, you may choose an example from your own experience.
4. Tell a dramatic story related to the subject, perhaps something you read in the press, for example, 'How the clothes of an operative became entangled in the moving parts of plant, causing serious injury, when he accidentally knocked the 'Start' button during servicing'.
5. Give some relevant statistics.

## Why – What incentive is there for people to listen?

1. Consider why anyone should listen. They won't listen simply because you are the foreman, supervisor or manager. You have to make it interesting and show concern for their health and safety.
2. Think about ways to make it more interesting or interactive, especially if the topic is well known and the talk is a reminder. People are generally more likely to retain information, and thus help them become safer, if they have actively taken part in something. So, for example, if it's a new type of PPE get them to try it on and ask for feedback.

3. A toolbox talk shouldn't be just **one way** communication. It is important to engage people in discussion, especially when talking to experienced people.

**Outline – It is important to explain here the major stages of your talk**

1. Clearly indicate the main areas you will concentrate on.
2. Decide on the key points you want to make and choose points that are relevant to your site situation.
3. People will only be able to cope with four or five pieces of information in each stage, so concentrate only on the relevant information.

## Site supervision simplified

*Toolbox talks* (GT 700) has been designed to follow the same structure and chapter layout as *Site supervision simplified* (GE 706). This will help you when preparing for a toolbox talk.

For example, if you want to deliver a talk on concrete and silica (B14 Silica dust), further information to prepare for the talk can be found within GE 706, Section B Health and Welfare, 10 Dust and fumes (Respiratory risks).

GE 706 is the official reference material for CITB's Site Safety Plus Site Supervisors' Safety Training Scheme (SSSTS), a two-day course for first line managers and supervisors.

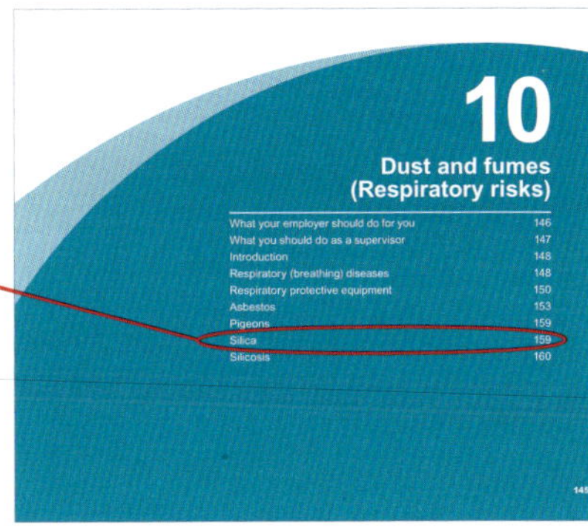
10

Dust and fumes (Respiratory risks)

| | |
|---|---|
| What your employer should do for you | 146 |
| What you should do as a supervisor | 147 |
| Introduction | 148 |
| Respiratory (breathing) diseases | 148 |
| Respiratory protective equipment | 150 |
| Asbestos | 153 |
| Pigeons | 159 |
| Silica | 159 |
| Silicosis | 160 |

145

## Toolbox talks CD-ROM

*The Toolbox talks CD-ROM* (GT 700/14 CD) contains all of the toolbox talks plus:

- ☑ a briefing record, delivery log and feedback form, which can all be printed, as required
- ☑ a presentation that will help trainers deliver an effective toolbox talk presenter's course to managers and supervisors in the building, construction and civil engineering industries, who will in turn be delivering toolbox talks to the workforce
- ☑ films of three toolbox talks being delivered by site personnel on a live site. Watching these talks being prepared and delivered in different styles will help you gain the confidence you need to prepare and deliver your own talks.

## Augmented reality

### Augmented reality – how to install the app

- ☑ Go to the appropriate app store (Apple or Android) and download the Layar app (free of charge) to your mobile phone or tablet.
- ☑ Look out for this logo [right], which indicates that you are on a Layar-friendly page.
- ☑ Open the Layar app and scan the Layar-friendly page.
- ☑ Wait for the page to activate on your device.
- ☑ Touch one of the buttons that have appeared to access additional content.

# A

# Legal and management

# Legal duties – What they mean to you

| | |
|---|---|
| **Reason** | You should be aware of your legal duties. |
| **Why** | You cannot comply with the law if you are not aware of what it says. |
| **Outline** | This talk explains the legal duties of employees under the Health and Safety at Work Act 1974 and regulations. |

# Legal duties – What they mean to you

| | |
|---|---|
| **Reason** | You should be aware of your legal duties. |
| **Why** | You cannot comply with the law if you are not aware of what it says. |
| **Outline** | This talk explains the legal duties of employees under the Health and Safety at Work Act 1974 and regulations. |

## Framework of health and safety law

1. The Health and Safety at Work Act 1974 provides general guidelines on the way in which work activities are to be carried out.
2. More detailed guidance is provided through the issue of regulations, which also carry the full force of law.

## Your legal duties under the Health and Safety at Work Act

1. You must safeguard your own health and safety and that of others (such as other workers and members of the public) who may be affected by your actions.
2. You must co-operate with your employer to help them comply with their legal duties.
3. You must not interfere with anything provided for health and safety.

## Your legal duties under the regulations

Some of the legal duties imposed on employees by regulations are shown below.

- ☑ **General safety.** To follow the training and instructions provided when using machinery, equipment, dangerous substances, transport equipment or safety devices.

  Report any defects which you believe could endanger health or safety.
- ☑ **PPE.** You must use personal protective equipment (PPE) in accordance with training and instructions given. Report loss or damage and store correctly after use.
- ☑ **Control of Substances Hazardous to Health (COSHH).** You must make use of any control measures provided to enable you to avoid contact with hazardous substances.
- ☑ **Noise.** You must wear hearing protection devices and take other actions that your employer may decide are necessary to protect your hearing.
- ☑ **Manual handling.** You must make use of any system of work provided by your employer to eliminate or reduce the risk of manual handling injuries.
- ☑ **Electricity.** You must co-operate with your employer and follow instructions with regard to working safely.

?

Give two examples of interfering with items provided for health and safety.
Describe one way in which you can assist your employer to comply with their legal duties.
What should you do if your safety helmet is damaged?

Encourage a discussion by using a real-life situation or example and ask if there are any questions.

# CDM Regulations – What they mean to you

| | |
|---|---|
| **Reason** | The CDM Regulations are the most important regulations for construction workers. |
| **Why** | If you know what your employer and you should do, you will be safe at work. |
| **Outline** | This talk covers some of the main topics in the day-to-day running of a safe site. |

*CDM highlights the importance of communication across roles*

*Example of a safety poster*

By having a questioning attitude you might make life a lot safer for everyone.

# CDM Regulations – What they mean to you

| | |
|---|---|
| **Reason** | The CDM Regulations are the most important regulations for construction workers. |
| **Why** | If you know what your employer and you should do, you will be safe at work. |
| **Outline** | This talk covers some of the main topics in the day-to-day running of a safe site. |

## Employers' duties

1. To act as a principal, main or sub-contractor and plan the work, put in place safe systems of work and explain how the work should be undertaken safely.
2. Provide and explain relevant paperwork (method statements and risk assessments) to supervisors and the workforce.
3. Ensure hazards have been removed from designs for work systems and methods of work, whenever possible.
4. Assess the competence of the work team and provide adequate training and supervision, where required.
5. Provide suitable welfare and security facilities for the site.
6. Provide the right information, to the right people, at the right time.

## Specific site duties

1. Workplaces, excavations, scaffolds, ladders, mobile elevating work platforms (MEWPs), and so on, must be suitable and checked by the user before use.
2. All plant must be in good working order, and be the right tool for the job.
3. Work activities must be programmed to minimise the risk of something going wrong.
4. Materials must be delivered, stored and used safely, taking into account manual and mechanical handling and any COSHH implications.

## Your duties

1. Follow instructions from your supervisor and make sure that you work safely.
2. Report anything that you think is unsafe. Stop work if you are not sure and ask for advice.
3. Know the site rules, including the emergency procedures.
4. Look after your tools, PPE and, most importantly, yourself and your workmates.

When you are given a job to do what do you expect from your supervisor?
What should you do if you find a scaffold with no visible signs of an inspection?
What should you do if you are unsure of your ability to carry out your task safely?
What should you do if you see one of your workmates doing something that you think is unsafe?

Encourage a discussion by using a real-life situation or example and ask if there are any questions.

# Company health and safety policies – What they mean to you

| | |
|---|---|
| **Reason** | Everyone at work must be aware of their company health and safety policy. |
| **Why** | We all need to understand our duties to protect our wellbeing and safety. |
| **Outline** | This talk covers some important aspects of both the employer's and your duties. |

*Make sure you understand your duties with regard to health and safety*

| **Notes** |
|---|
| |
| |
| |
| |
| |
| |
| |

When you look in a mirror who do you see?
The answer should be easy – your own health and safety person facing you.

# Company health and safety policies – What they mean to you

| | |
|---|---|
| **Reason** | Everyone at work must be aware of their company health and safety policy. |
| **Why** | We all need to understand our duties to protect our wellbeing and safety. |
| **Outline** | This talk covers some important aspects of both the employer's and your duties. |

## Company health and safety policy

1. A company health and safety policy is required by law and is a written statement of how your employer manages health and safety.
2. Everyone has duties and is provided with protection by following the guidance set out in the health and safety policy.
3. Legal powers of enforcement are placed with the Health and Safety Executive (HSE) for construction and, in some cases, the Local Authority's Environmental Health Department.
4. Everyone has a health and safety responsibility, both for themselves and others who may be affected by acts or omissions.

## Health and safety policy content

1. The health and safety policy contains information on how a company should manage its legal duties to comply with the law.
2. A statement of intent or policy at the beginning of the document sets out the intentions of the company and how it will manage its business in order to comply with legal requirements.
3. This is followed by details of the organisation's structure and the arrangements for policy implementation.
4. Also included are the duties of the individual duty holder, which includes everyone.

## What it means for you

1. You should have access to a copy of your employer's health and safety policy, understand its contents and follow guidance set out in it.
2. The policy is a set of rules to protect the workforce and others.
3. On average 82 workers and 181 members of the public are killed every year by work activities. The construction industry has the biggest share of these incidents.
4. If you have a doubt or concern you have a legal duty to ask for an explanation.
5. We are all our own safety supervisors – take care of yourself and others!

## Consultation

1. Consultation is more than people just giving you information, it is about employers listening and taking account of what you say, before they make decisions that will affect your health and safety.
2. Employers can carry out consultation by:
   - talking to and listening to trade union safety representatives or other appointed representatives
   - having regular health and safety committees or forums
   - using inductions, daily briefings and toolbox talks to explain what is happening, and to listen to and act upon your comments
   - talking to you directly during informal visits walkabouts
   - setting up a system that lets you report problems or suggest safer ways of doing your work.
3. These measures should make construction work safer so that you can return to your family at the end of the day.

When should you ask questions about the company health and safety policy?
Why is it important to report unsafe situations?
If you found a workmate doing something unsafely what immediate actions should you take?
What actions could you take to improve your safety in your workplace?

Encourage a discussion by using a real-life situation or example and ask if there are any questions.

# Personal competence

| | |
|---|---|
| **Reason** | Competence of people is important for the safety of everyone at work. |
| **Why** | It is important that everyone knows the limits of their personal competence. |
| **Outline** | This talk explains what is meant by competence of both the employer and individuals. |

*Examples of competency card schemes*

*Competence covers several aspects*

# Personal competence

| | |
|---|---|
| **Reason** | Competence of people is important for the safety of everyone at work. |
| **Why** | It is important that everyone knows the limits of their personal competence. |
| **Outline** | This talk explains what is meant by competence of both the employer and individuals. |

## Employer competence

1. Employer competence is sometimes known as corporate competence.
2. It is the ability of the company to manage health and safety through its policies and procedures (setting targets).
3. Company competence (your manager or supervisor) provides you with a safe place of work, with safe access and egress, method statements and risk assessments, along with the correct tools needed to do your job.
4. The company expects you to have a questioning attitude and think proactively, before taking any action, to ensure the action is appropriate and safe.

## Personal competence

1. Your competence is the key to a successful and safe business and the company relies on you to make it happen.
2. Personal competence is a blend of:
   **skills** – your ability to manage and influence others
   **knowledge** of health and safety issues
   **attitude** – your wish to achieve your health and safety aims
   **training** to gain the necessary knowledge and/or qualifications
   **experience**, which can only be achieved over a period of time.
3. Using your personal competence can ensure not only your safety but also that of others.
4. It provides the route for a successful project. A project that is well carried out and meets all of the necessary targets.

## Benefits of personal competence

1. You can achieve job satisfaction and respect for the quality of the work that you produce.
2. You will be relied upon for your expertise and may have the potential for promotion.
3. A competent worker is a safe worker.

When should you question the work method you have been asked to use?
If the method statement and risk assessment don't match the task what should you do?
What are your duties with regard to your workmates?
If you thought you needed more help or training what action should you take?

Now inform your workforce of the company provision for training.
Encourage a discussion by using a real-life situation or example and ask if there are any questions.

# Risk assessments and method statements

| | |
|---|---|
| **Reason** | Work must be planned and carried out in a safe manner. |
| **Why** | The construction industry continues to have an unacceptable accident record. |
| **Outline** | This talk covers what risk assessments and method statements mean to you. |

| | | | | |
|---|---|---|---|---|
| **Likelihood** | **High** | **High** | **High** | **High** |
| | **Medium** | **Medium** | **Medium** | **High** |
| | **Low** | **Low** | **Medium** | **High** |
| | | **Low** | **Medium** | **High** |
| | | **Severity** | | |

*Risk assessments can prevent accidents*

# Risk assessments and method statements

| | |
|---|---|
| **Reason** | Work must be planned and carried out in a safe manner. |
| **Why** | The construction industry continues to have an unacceptable accident record. |
| **Outline** | This talk covers what risk assessments and method statements mean to you. |

## Risk assessments

1. All employers have a legal duty to prepare risk assessments for work activities that could foreseeably result in injury to any person or damage to equipment.
2. Risk assessments outline the ways in which the job could result in injury or damage and the measures that must be put in place to ensure that the chance of anything going wrong is eliminated or reduced to an acceptable level.
3. Employers with five or more employees must have written risk assessments.
4. If there are fewer than five employees, the risk assessments must still be carried out, although there is no legal duty to write them down.
5. Employers have a legal duty to communicate the significant findings of risk assessments to workers who may be affected.
6. There is no specified way for laying out a risk assessment so you must familiarise yourself with the way your employers present theirs.
7. In many cases, the risk assessments will form the basis for a method statement.

## Method statements

1. Method statements are a written list of operations, to be carried out in a specified sequence, in order to complete a work activity in a safe manner.
2. Everyone involved in a job for which a method statement has been written, should read it (or have it explained to them) and sign it as having understood its contents.
3. Well-written method statements address all the hazards present and plan the work so that the risk of accident is eliminated or reduced to an acceptable level.
4. Method statements should be site specific.

What would you expect to find in a risk assessment?
If your company has five or more employees, how would its risk assessments be presented?
What should happen if you need to deviate from the method statement?

Encourage a discussion by using a real-life situation or example and ask if there are any questions.

# Young people on site

| | |
|---|---|
| **Reason** | Construction sites are hazardous places, even for adults who are aware of the dangers. |
| **Why** | Young people, with their lack of safety awareness, are particularly at risk of work-related injury or ill health. |
| **Outline** | This talk covers why young people are vulnerable and what everyone should be doing to safeguard their health and safety on site. |

*Make sure young people understand the importance of site safety*

*Young people need more supervision than adults*

# Young people on site

| | |
|---|---|
| **Reason** | Construction sites are hazardous places, even for adults who are aware of the dangers. |
| **Why** | Young people, with their lack of safety awareness, are particularly at risk of work-related injury or ill health. |
| **Outline** | This talk covers why young people are vulnerable and what everyone should be doing to safeguard their health and safety on site. |

## Who is a young person?

1. Health and safety law defines a young person as anyone under 18.
2. The law does not prohibit the employment of young people on construction sites.

## What are the problems?

1. Young people will not have the same level of safety awareness as a more experienced person.
2. For some young people a construction site will be their first experience of a workplace.
3. Changes in legislation have removed the minimum age limit of 18 for operating plant and lifting equipment; but young people may not appreciate their own limitations.
4. Young people will require a greater level of supervision than an adult, the level depending on their job and site conditions.
5. Young people might be more tempted to arrive at work whilst unfit, due to several factors.
6. Young people might create dangerous situations because of an eagerness to please.

## Protection of young people

1. Be aware of their lack of safety awareness, their physical and psychological immaturity and their inexperience.
2. Risk assessments must take account of young people on site and jobs that they are required to do.
3. Only give jobs to young people that they can cope with, both physically and mentally.
4. Do not allow young people to carry out particularly dangerous jobs (such as using cartridge-operated tools).
5. Whatever the job, you must ensure that the level of supervision is adequate.
6. If young people are working near to you, you should be ready to stop them if they carry out any activity that is clearly unsafe.
7. Encourage young people to speak out if they do not feel safe with what they have been asked to do; it may only be a case of reassurance or maybe more supervision is required.
8. Ensure that young people attend site induction, even if they are only going to be on site for a short time.
9. Do not tolerate messing about, or other unsafe actions or high spirits.

What action should you take if you witness a young person in a position of danger?
What do you do if a young person arrives unfit for work?
How do you assess the correct level of supervision?

Now inform your workforce of the company policy regarding safeguarding the health and safety of young people on site.
Encourage a discussion by using a real-life situation or example and ask if there are any questions.

# Worker engagement

| | |
|---|---|
| **Reason** | Involving the workforce in management decisions can help to improve health and safety performance. |
| **Why** | The workforce is an important part of every business and engaging with them is a key element to success. |
| **Outline** | This talk covers the key principles required in the relationship between the employer and you. |

*Staff in all roles should work together to ensure safe practices*

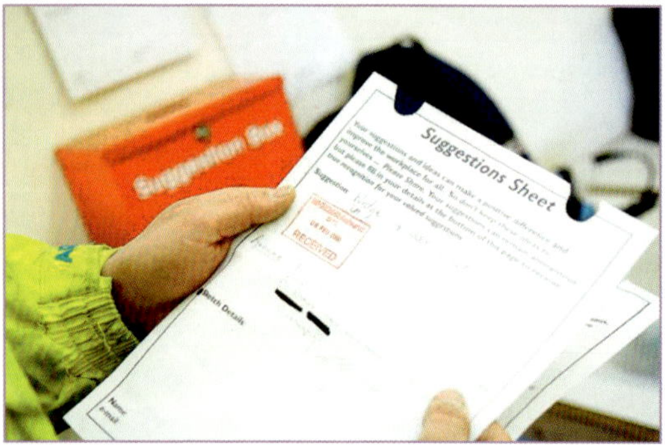


*Make suggestions to improve your site*

If you see something that is not right, report it. Today's near miss is tomorrow's accident waiting to happen.

# Worker engagement

| | |
|---|---|
| **Reason** | Involving the workforce in management decisions can help to improve health and safety performance. |
| **Why** | The workforce is an important part of every business and engaging with them is a key element to success. |
| **Outline** | This talk covers the key principles required in the relationship between the employer and you. |

## Involving the workforce (worker engagement)

1. All employers have a legal duty to involve you in decisions made with regard to health and safety matters.
2. People doing the job frequently have a more accurate idea of how to improve not only health and safety but also performance.
3. Communication, co-operation and establishing trust are important elements in worker engagement.

## Engaging with you

1. You should be encouraged to **stop** work and seek advice when you feel unsafe.
2. If you have any concerns, or think you can make your work or workplace safer, then tell your supervisor.
3. Treat others (such as sub-contractors) in the same way as you would your own workmates.
4. Be prepared to be asked for your opinion and contribute if you can.

## Benefits for everyone

1. By having an open, blame-free culture, work will be carried out more safely and efficiently.
2. Mistakes and near misses can be identified and control measures put in place to prevent recurrence.
3. Open communication, where good and bad things are freely discussed, has benefits for both production and safety.

Why is it important to be able to approach your supervisor about a problem?
What should you do if you think you can make a job safer?
If you do not get help from your immediate supervisor what should you do next?
How important is it that you are involved in decisions at work?
Do you have any suggestions on how to improve the way you are working?

Now inform your workforce of the company or site policy on worker engagement.
Encourage a discussion by using a real-life situation or example and ask if there are any questions.

# Fairness, inclusion and respect – Your responsibilities

| | |
|---|---|
| **Reason** | Employers and employees need to be aware of their rights and responsibilities. |
| **Why** | Workplaces are more productive and healthy in an environment free from bullying and harassment. |
| **Outline** | This talk covers legal and contractual rights and responsibilities covering dignity in the workplace, and will raise awareness of the issues that can affect good working relationships. |

| **Notes** |
|---|
| |
| |
| |
| |
| |
| |
| |
| |
| |
| |
| |
| |
| |
| |
| |
| |
| |
| |
| |
| |
| |
| |
| |
| |
| |

# Fairness, inclusion and respect – Your responsibilities

| | |
|---|---|
| **Reason** | Employers and employees need to be aware of their rights and responsibilities. |
| **Why** | Workplaces are more productive and healthy in an environment free from bullying and harassment. |
| **Outline** | This talk covers legal and contractual rights and responsibilities covering dignity in the workplace, and will raise awareness of the issues that can affect good working relationships. |

## What the law says

1. Everyone in the construction industry has legal duties under the Equality Act.
2. Companies will have rules of behaviour within their own terms and conditions, some of which may extend beyond the protection afforded within the Equality Act.
3. Everyone in the construction industry is protected by the Act and their company rules of behaviour.
4. The Act supports and protects any worker challenging behaviours, even when not directed at them.
5. The Act puts responsibilities on managers to act swiftly to challenge behaviours.
6. The Act gives you the right not to be harassed by third parties and makes the employer potentially liable for harassment of their employees, by people such as customers or clients. (The employer will only be liable when harassment has occurred on at least two previous occasions, they are aware that it has taken place and have not taken reasonable steps to prevent it from happening again.)

## Provisions of the Act concerning harassment

1. Under the Act your employer has a legal duty to provide a workplace free from bullying and harassing behaviour. They should:
   - make it clear in their terms and conditions, training, or policies and procedures that such behaviour will not be tolerated
   - take any complaint seriously
   - challenge your behaviour to safeguard others who may be affected by your actions and behaviours
   - ensure breaches in behaviour are investigated and treated in line with disciplinary procedures (or dignity at work procedures).
2. Under the Act you, as an employee, have a legal duty to ensure you treat co-workers, clients and the public with dignity and respect.

## Other provisions

1. **Unfair dismissal.** Employers have a duty of care for all their employees. If the mutual trust and confidence between employer and employee is broken (for example, through bullying and harassment at work) then an employee can resign and claim constructive dismissal on the grounds of breach of contract. Employers are usually responsible in law for the acts of their workforce.
2. **Health and safety breach of contract.** Under the Health and Safety at Work Act 1974 employers are responsible for the health, safety and welfare at work of all employees.

## What else can you do?

1. Unfortunately, there may occasionally be times when not all employers or managers follow the correct procedures. In these situations, there are organisations you can contact for support, advice or guidance.

   **Advisory, Conciliation and Arbitration Service (ACAS) – Helpline 08457 474747**

   ACAS is an independent body with statutory duties. It provides information, advice and training, and works with employers and employees to solve problems and improve performance in the workplace.

Who has responsibility for ensuring the workplace is free from harassment and everyone is treated fairly and with respect?

Which legislation protects you from behaviour affecting your dignity and respect?

Encourage a discussion by using a real-life situation or example and ask if there are any questions.

# Fairness, inclusion and respect – Health, work and wellbeing

| | |
|---|---|
| **Reason** | Everyone needs to be aware of their legal responsibilities (Equality Act) regarding disability and health, and of common health problems and supportive behaviours and actions to promote wellbeing at work. |
| **Why** | There is a higher risk of occupational disability within construction than in many other industries. The company have a duty to make reasonable adjustments to accommodate the change required for someone who is either injured at work or develops a health issue that impacts on their ability to undertake their full role. Therefore, when people with a disability return to the workplace, it is important to understand the legal framework that protects disabled workers, as well as the actions and behaviours that support an inclusive, healthy and productive workplace. |
| **Outline** | This talk covers the Equality Act and disabilities, common health problems and the link between health and wellbeing and work. |

## Stress

1. The Health and Safety Executive (HSE) define stress as the *adverse reaction people have to excessive pressures or other types of demand placed on them.*
2. Stress has been associated with anxiety and depression, heart disease, back pain and gastrointestinal illnesses.
3. Health and safety law expects an employer to assess and take measures to control risks from work-related stress.
4. You are responsible for talking to your manager about any factors outside the workplace that can turn otherwise normal pressures of work into excessive ones.
5. In construction, stress could be caused by many triggers, especially the time constraints of projects and the need to get the job done. It can lead to you feeling more angry or being more tired.
6. Stress and time pressures could also lead to you taking more risks with your long-term health (such as a two-minute job that may release silica dust and therefore requires you to wear a face mask, can easily turn into a 15-minute job if you have to go to the store to collect the necessary personal protective equipment).

## Other issues affecting health and wellbeing

1. **Drug and alcohol abuse.** Companies should have clear guidelines on this.
2. **Hazardous substances.** The HSE produces guidance on how to manage hazards posed by chemical and dangerous substances.
3. **Bullying and harassment.** Companies should have clear guidelines on this as it can severely damage health and wellbeing. Harassment is unlawful under the Equality Act.

Who has responsibility for your health and wellbeing at work?
What examples can you give of disabilities covered by the Equality Act?
What reasonable adjustments could be made?
How can mental health be managed?
What is stress and who is responsible for managing it in the workplace?
What other issues can affect your health and wellbeing at work?

Encourage a discussion by using a real-life situation or example and ask if there are any questions.

# Fairness, inclusion and respect – Health, work and wellbeing

| | |
|---|---|
| **Reason** | Everyone needs to be aware of their legal responsibilities (Equality Act) regarding disability and health, and of common health problems and supportive behaviours and actions to promote wellbeing at work. |
| **Why** | There is a higher risk of occupational disability within construction than in many other industries. The company have a duty to make reasonable adjustments to accommodate the change required for someone who is either injured at work or develops a health issue that impacts on their ability to undertake their full role. Therefore, when people with a disability return to the workplace, it is important to understand the legal framework that protects disabled workers, as well as the actions and behaviours that support an inclusive, healthy and productive workplace. |
| **Outline** | This talk covers the Equality Act and disabilities, common health problems and the link between health and wellbeing and work. |

## Scope of the Equality Act – Disability

1. Everyone in the company is protected by the Equality Act and has legal duties regarding disability under the legislation.
2. Other provisions under health and safety legislation set out duties to control work-related health risks.

## General provisions of the Equality Act for disabilities

1. A disability is a physical or mental impairment, which has a substantial and long-term adverse effect on a person's ability to carry out normal, day-to-day activities.
2. A person with a disability at work is protected from discrimination. This means that employers must not treat a person with a disability less favourably because of their disability, without a justifiable reason.
3. Employers are required to make reasonable adjustments to working conditions or the workplace where that would help to accommodate a particular person with a disability.
4. The Act includes a new protection from discrimination arising from disability. It is discrimination to treat a person with a disability unfavourably because of something connected with their disability (such as a tendency to make spelling mistakes arising from dyslexia) or discrimination by association (such as if someone has caring responsibilities for a person with a disability).
5. One in five people in the UK has a disability.

## Connection between health and work

1. The connection between physical hazards (such as noise, dust and chemical hazards) and health are well recognised amongst employers and employees. Less understood is the relationship between mental and physical wellbeing and the jobs we do.
2. Over two million people have reported suffering from an illness (such as stress, anxiety, back pain, depression or increased risk of coronary heart disease) that they believe has been caused, or made worse, by their work.

## Mental health

1. Every year three in ten employees experience mental health problems.
2. By 2020 depression will rank second to heart disease as the leading cause of disability worldwide.
3. Mental health can affect anyone, regardless of their age, gender, ethnicity or social group. The most common forms of mental ill health are anxiety, depression, phobic anxiety disorders and obsessive compulsive disorders.
4. It is classed as a disability under the Equality Act, which makes it unlawful for an employer to treat anyone with a mental health problem less favourably, without a justifiable reason.

# Fairness, inclusion and respect – Respect

| | |
|---|---|
| **Reason** | Everyone needs to be aware of what respect in the workplace means. |
| **Why** | Workplaces are more productive and healthy in an environment free from bullying and harassment. |
| **Outline** | This talk covers legal and contractual rights and responsibilities covering dignity in the workplace, and will raise awareness of the issues that can affect good working relationships. |

## Types of bullying

1. Bullying takes on many different forms.
   - **Physical bullying** includes any physical contact that would hurt or injure a person.
   - **Verbal bullying**, including name-calling or making offensive remarks or jokes.
   - **Indirect bullying** includes spreading rumours or stories about someone, telling others about something that was told to you in private and excluding others from groups.
   - **Social exclusion**, where victims are deliberately excluded from group activity.
   - **Intimidation**, where someone is pressured into doing what the bully wants against what they believe to be right.
   - **Cyber-bullying** uses e-technology as a means of victimising others. It is the use of an internet service or mobile technologies (such as email, chat-room discussion groups, social networking, instant messaging, web pages or SMS (text messaging)) with the intention of harming another person.
2. It is also bullying if you feel hurt because of things said about your ethnic background, religious faith or beliefs, sex, sexuality, disability, educational needs, appearance or family issues.

## Effects of bullying and harassment

1. **On the recipient** can include low self-esteem, antisocial behaviour, illness (such as stress, rashes and insomnia), reluctance to return to work, absenteeism, anxiety, depression and possible suicidal tendencies.
2. **At the workplace** can include time wasted dealing with negative behaviour and absenteeism, can affect winning future jobs or give a negative reputation.
3. The effects of bullying could give a bad reputation for potential new workers and future customers.

?

What examples can you give of bullying or harassing behaviour that you have seen or experienced?
What other characteristics might be covered by company policies and disciplinary procedures?
What should you do if you are being harassed?
Why do you think it is important to challenge bullying and harassing behaviour?
What do you think it would feel like to be on the receiving end of bullying and/or harassing behaviour?

Encourage a discussion by using a real-life situation or example and ask if there are any questions.

# Fairness, inclusion and respect – Respect

| | |
|---|---|
| **Reason** | Everyone needs to be aware of what respect in the workplace means. |
| **Why** | Workplaces are more productive and healthy in an environment free from bullying and harassment. |
| **Outline** | This talk covers legal and contractual rights and responsibilities covering dignity in the workplace, and will raise awareness of the issues that can affect good working relationships. |

## What does respect mean?

1. Respect is ensuring your behaviour towards teams, colleagues, clients or contactors is appropriate and does not cause offence.
2. It is about treating people how you would expect to be treated and maintaining an environment where individual differences are respected.
3. Most companies will have a policy setting out these standards of behaviour and will link them to their discipline and grievance procedures. Some are referred to as dignity and respect, others may be known as anti-bullying and harassment policies.

## What are harassment and bullying?

1. **Harassment** is unwanted conduct affecting the dignity of men and women – any actions or comments that are viewed as demeaning and unacceptable to the recipient.
2. **Bullying** is the deliberate action or behaviour directed towards another person that has the effect of causing pain and distress to the victim.
3. Bullying or harassment can take many forms. It may be by a one-to-one situation or involve groups of people. Whatever the form of harassment, it will be unwanted behaviour that is unwelcome and unpleasant.

## Harassment and the Equality Act – Protected characteristics

1. Most company anti-bullying and harassment policies apply to any form of harassment, whether related to a personal characteristic or not, and penalties for any breaches of the policy can be imposed internally through the company's disciplinary procedures.
2. Harassment is unlawful under the Equality Act and applies to the protected characteristics (age, disability, gender reassignment, race, religion or belief, sex and sexual orientation). Harassment complaints related to these characteristics can be heard at an employment tribunal, and any that are upheld can lead to the individual being held personally liable, as well as the company.
3. The Equality Act does not specifically prohibit harassment relating to pregnancy and maternity or marriage and civil partnership. However, individuals subjected to detriment as a result of pregnancy and maternity or marriage and civil partnership may be protected through other protected characteristics (such as sex or sexual orientation).
4. The Equality Act allows employees to complain of behaviour that they find offensive, even if it is not directed at them, and the complainant need not possess the relevant personal characteristic themselves.
5. Sexual harassment is one of the most common forms of harassment and is specifically outlawed by the Equality Act.
6. Employees are also protected from harassment because of perception (if others think someone possesses a particular protected characteristic) and association (if they are associated with someone who has a protected characteristic (such as a family member or friend)).

# Fairness, inclusion and respect – Acceptable language

| | |
|---|---|
| **Reason** | Everyone needs to be aware of what language is acceptable when working on site. |
| **Why** | Workplaces are more productive and healthy when workers, staff and visitors feel safe and not threatened. |
| **Outline** | This talk covers what is inappropriate language and why it is disrespectful, and it will raise awareness of issues that can affect good working relationships. |

## What has this got to do with fairness, inclusion and respect?

1. The underpinning foundation of fairness, inclusion and respect is that everyone is treated with respect and consideration and is not threatened or intimidated by the words or actions of others.
2. There is not a sector of the community that will not have members that will feel at least one of those emotions when hearing inappropriate language.
3. People are protected by law from this type of harassment. In particular, any language used to put down and disrespect them (or others) because of, for example, age, race, religion, sex or sexual orientation.
4. Managers are instructed to challenge the use of bad language by workers and it could result in disciplinary action.

?

What examples can you give of inappropriate language that you have seen or experienced?
What should you do if you are being harassed?
Why do you think it is important to challenge bullying and harassing behaviour?
What do you think it may feel like to be on the receiving end of inappropriate banter?
What do you think it may feel like to be on the receiving end of bullying and/or harassing behaviour?

Encourage a discussion by using a real-life situation or example and ask if there are any questions.

# Fairness, inclusion and respect – Acceptable language

| | |
|---|---|
| **Reason** | Everyone needs to be aware of what language is acceptable when working on site. |
| **Why** | Workplaces are more productive and healthy when workers, staff and visitors feel safe and not threatened. |
| **Outline** | This talk covers what is inappropriate language and why it is disrespectful, and it will raise awareness of issues that can affect good working relationships. |

## What is inappropriate and disrespectful language?

1. Workers using bad and/or abusive language in places where they can be clearly heard by people nearby is unacceptable and consideration needs to be given to the environment and those around you.
2. Language evolves and individuals may consider their use of expletives as an acceptable part of everyday conversation. In some cases, between two like-minded people, it could be viewed as acceptable.
3. Banter is a way of communication on many sites and good banter can make a hard job enjoyable. However, when banter is aimed at a particular person's personal characteristic it can lead to them feeling excluded from the group. Comments made by one individual once a day can be repeated by many individuals over the course of the day and that banter then turns from a fun exchange of comments to a negative aspect of work for some people.

## Why do you use this type of language?

1. To appear more adult and streetwise and so gain respect from other workers?
2. You always use it and it has nothing to do with anyone else?
3. You cannot express yourself without using such language?
4. It is a habit you have got into?
5. You use it at home and so can't see why you shouldn't use it elsewhere?

## What effect will it have on others?

1. It may make them less likely to want to return.
2. You could intimidate the person who hears it or make them feel uncomfortable.
3. A person who was considering working on site may change their mind.
4. It could have a negative effect on young or impressionable people (such as if you are working in a community or at a school).

## What effect will it have on your employer?

1. Your organisation could look unprofessional as a result.
2. It might put people off from employing your company.
3. Clients or partner companies you are working with might become upset and it could then affect repeat business.

# Checks and inspections

| | |
|---|---|
| **Reason** | Checks, inspections and record keeping are an important part of site safety. |
| **Why** | Construction site equipment can be exposed to harsh working conditions, but checks help to ensure it is safe to use or can highlight a fault that requires fixing. |
| **Outline** | This talk covers the types of checks and inspections that should be carried out. |

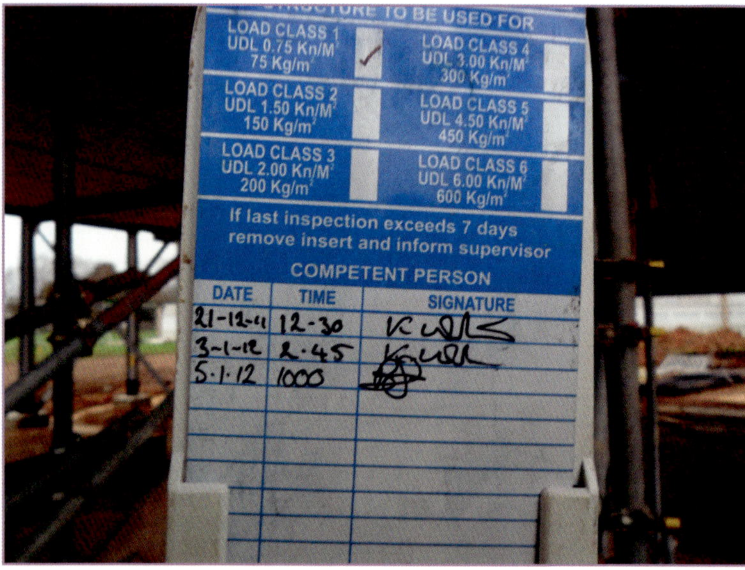


*Check the scaffold has been inspected before you access*

| Notes |
|---|
| |
| |
| |
| |
| |
| |
| |

If you notice unsafe tools, plant, access systems (such as ladders or scaffolds) or excavations that look unstable, **stop work and report it to your supervisor.**

# Checks and inspections

| | |
|---|---|
| **Reason** | Checks, inspections and record keeping are an important part of site safety. |
| **Why** | Construction site equipment can be exposed to harsh working conditions, but checks help to ensure it is safe to use or can highlight a fault that requires fixing. |
| **Outline** | This talk covers the types of checks and inspections that should be carried out. |

## What the law says

1. Regulations place duties on employers to carry out formal statutory inspections on a weekly basis and for the findings to be recorded.
2. Scaffold, plant and tool tags are often used as a visible sign that equipment and work areas have been examined, but a computer or paper record must be completed and retained to comply with the law.
3. Weekly checks have to be carried out by a trained and competent person and they must also complete a record of their findings.
4. Everyone on site has a duty to carry out their own workplace and work tool checks.

## Frequency of checks and inspections

1. Tools and equipment (hand and power portable) should be checked daily by the user and defects (such as damaged plugs, sockets and leads) reported.
2. Checklists are provided with larger plant and equipment – ensure both the daily and weekly checks are carried out, as required.
3. Lifting equipment and accessories (such as cranes and slings) require daily, weekly, monthly and yearly inspections.
4. Access equipment (such as scaffolds and ladders, pop-ups and podiums) must be visually inspected every time they are used.
5. Excavations have to be inspected at the start of every shift and weekly records retained.

## Statutory examinations

1. On-site plant (such as electrical equipment and RCDs) must be portable appliance tested (PAT) every three months.
2. Mechanical plant (such as MEWPs), lifting accessories (such as chains and slings) and safety nets, as well as cranes used for lifting people, must be examined every six months.
3. Virtually everything else has to be examined every 12 months, including fire-fighting appliances, site office electrical systems and equipment, cranes and tower cranes.

?

When should you inspect your portable tools and equipment?
Why is it important to report any defects that you find?
What should you do if you notice an inspection tag date has expired?
What should you do if you are working on a scaffold and notice that a guard-rail is missing?

Now inform your workforce of the company provision for inspection and examination of plant and equipment.
Encourage a discussion by using a real-life situation or example and ask if there are any questions.

# Accident and incident reporting

| | |
|---|---|
| **Reason** | Accurate accident investigation and prompt accident reporting are important. |
| **Why** | Establishing why accidents occur and examining their effects can help prevent them from being repeated. |
| **Outline** | This talk covers reporting and recording procedures. |

**Unsafe conditions**

*Something with the potential to cause harm*

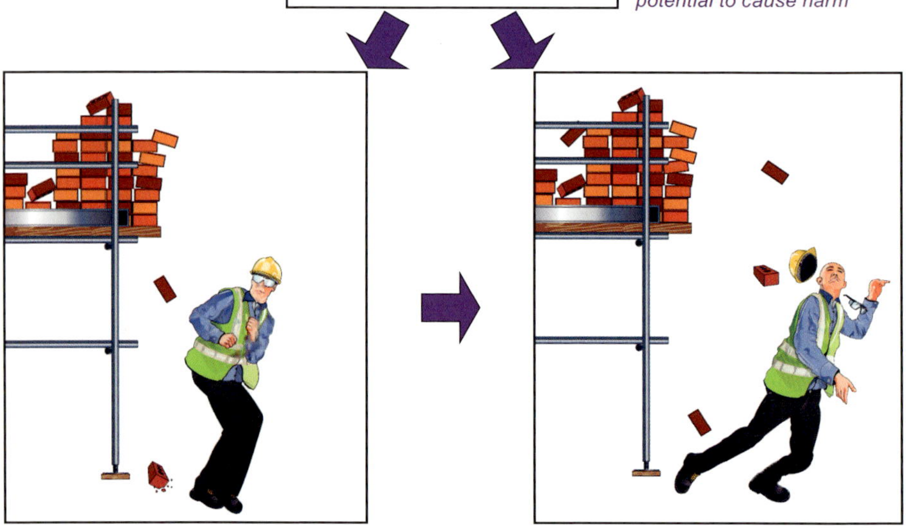

**Near misses**

*An incident that nearly resulted in an injury or damage*

**Accidents**

*An incident that resulted in an injury or damage*

Preventing an accident is always possible – mending broken lives and bodies is not.

# Accident and incident reporting

| | |
|---|---|
| **Reason** | Accurate accident investigation and prompt accident reporting are important. |
| **Why** | Establishing why accidents occur and examining their effects can help prevent them from being repeated. |
| **Outline** | This talk covers reporting and recording procedures. |

## Accident reporting

1. Health and safety law requires that the following types of accident are reported to the HSE:
   - fatalities and specified injuries
   - injuries resulting in more than seven days off work or inability to carry on with normal work
   - dangerous occurrences.
2. By receiving such accident reports the HSE and your company can establish accident trends, highlight areas of weakness and effectively target preventative measures.
3. Everyone on site must ensure that all accidents, no matter how minor, are recorded in the site accident book.
4. Completed accident reports are confidential.
5. If accurate records are made, the affected parties can refer back to them at a later date, if there is a need.
6. Accidents to members of the public arising out of site activities must be reported.

## Accident investigation

1. Your employer has a duty to investigate all accidents to establish the cause and prevent recurrence.
2. The HSE will also investigate fatalities and other serious accidents.
3. If you are involved in an investigation:
   - listen carefully to the questions and remain calm
   - state honestly what you saw or heard
   - do not be afraid to say when you do not know an answer.
4. Remember that the reason for the investigation is to prevent the accident happening again.

?

What action should you take if you witnessed an accident to another person?
Why is it important that all accidents are recorded in the accident book?
Who should be informed in your company if an accident has just occurred?
Why is it important to fully co-operate with someone who is carrying out an accident investigation?

Inform those present of the location of the accident book and the company procedure for recording and investigating accidents, including the requirement for employers to ensure that completed accident reports are kept confidential.
Encourage a discussion by using a real-life situation or example and ask if there are any questions.

# B
# Health and welfare

# Skin protection

| | |
|---|---|
| **Reason** | Dermatitis accounts for over half of all working days lost through industrial sickness. |
| **Why** | Some types of dermatitis are so severe sufferers have to give up work or change their trade. |
| **Outline** | This talk covers hazards to the skin and precautions to protect your skin. |

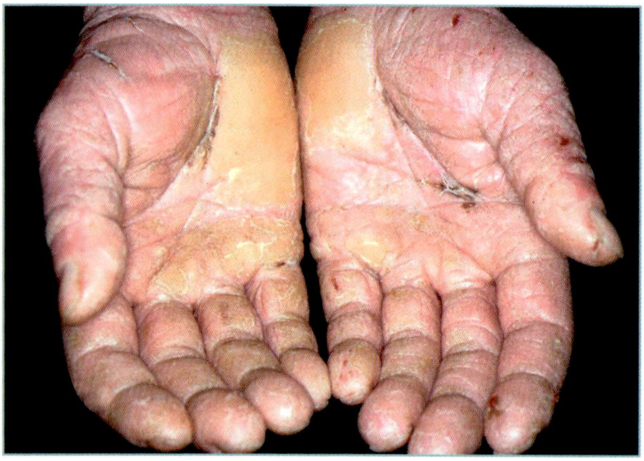

*Dermatitis showing crusting and thickening of skin*

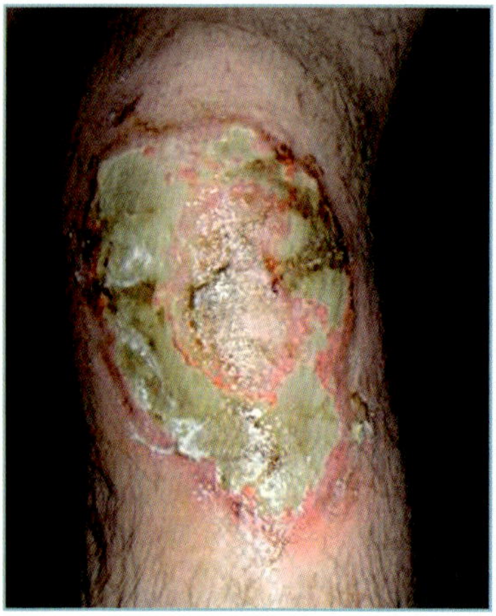

*Irritant contact dermatitis ('pizza knee') from a cement burn*

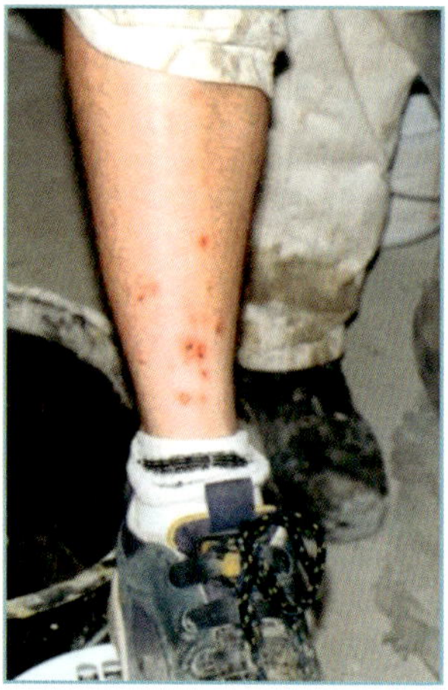

*Allergic contact dermatitis of the leg*

If you notice a rash or warts, see your family doctor at once.

# Skin protection

| | |
|---|---|
| **Reason** | Dermatitis accounts for over half of all working days lost through industrial sickness. |
| **Why** | Some types of dermatitis are so severe sufferers have to give up work or change their trade. |
| **Outline** | This talk covers hazards to the skin and precautions to protect your skin. |

## Contact hazards to skin

1. Mineral oils, including fuel oils and mould oils, can give you bad skin conditions, oil acne, or even cancer.
2. Prolonged skin contact with oily rags in overall pockets can cause testicular cancer.
3. Chemicals, including alkalis, acids and chromates can penetrate the skin causing ulcers and dermatitis.
4. Cement can cause chronic dermatitis. Wet cement can cause serious burns, which could lead to the amputation of a limb.
5. Solvents and degreasers (such as thinners) dissolve natural oils in skin leaving it open to infection.
6. Tar, pitch and bitumen products cause blisters and oil acne. They can also cause tar warts, leading to cancer.
7. Epoxy-resins, glass fibre and some hardwoods irritate the skin and can lead to dermatitis.
8. Extremes of sunshine, temperature and humidity make the skin more susceptible to dermatitis and other skin problems.

## Precautions to protect your skin

1. Comply with your employer's safe system of work.
2. Avoid skin contact with hazardous substances and wear the correct personal protective equipment.
3. Keep your skin clean and use after-wash skin cream.
4. Keep your workplace clean.
5. Get first aid for cuts and grazes and keep them covered.
6. Don't use abrasives or solvents to clean your skin.
7. Don't let synthetic resins or glue harden on your skin.
8. If you experience skin contact with hazardous substances, periodically examine the affected part of your skin for signs of damage.
9. Never wear oil-contaminated clothes next to your skin.
10. Seek professional medical advice if necessary.

What hazards are there from contact with mineral oil?
What effect do solvents have on your skin?
What can you wear to protect your skin?
If you notice rashes or warts what should you do?
What should your employer be doing for you?
What should not be used to clean your skin?
How can exposure to the sun affect you?

Now inform your workforce of the company policy regarding skin care.
Encourage a discussion by using a real-life situation or example and ask if there are any questions.

# Sun protection

| | |
|---|---|
| **Reason** | You should know the simple steps to take to protect yourself from the sun. |
| **Why** | Outdoor workers can experience excessive exposure to the sun's UV radiation and are therefore more at risk from skin cancer. |
| **Outline** | This talk covers the facts and statistics about skin cancer, who's at risk and how to protect yourself. |

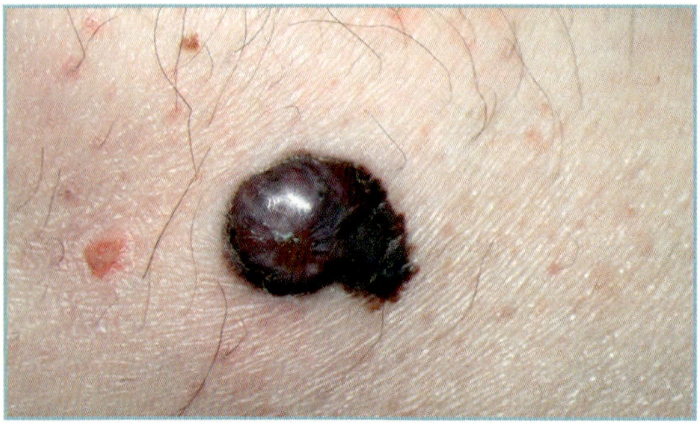

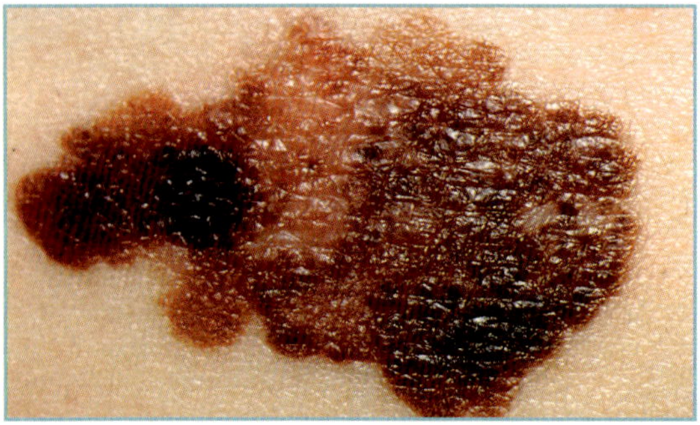

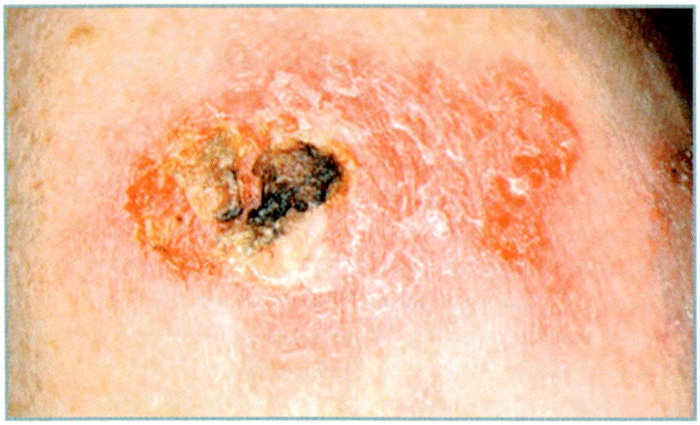

*See a doctor immediately if you are concerned about a mole*

Moles, such as these, should be checked by a doctor – don't put it off.

# Sun protection

| | |
|---|---|
| **Reason** | You should know the simple steps to take to protect yourself from the sun. |
| **Why** | Outdoor workers can experience excessive exposure to the sun's UV radiation and are therefore more at risk from skin cancer. |
| **Outline** | This talk covers the facts and statistics about skin cancer, who's at risk and how to protect yourself. |

## Facts and figures

1. UV (ultraviolet) radiation from the sun is a major cause of skin cancer. Cases have doubled in the last 20 years. There are two types of skin cancer – one is less common but more serious, while the other is more common but less serious. Approximately 95,000 people are diagnosed with skin cancer and 2,700 die from it each year.
2. Skins cancers are more common in females in the younger age groups but the pattern reverses in older people.
3. Sunlight causes the skin to produce a dark pigment called melanin; this is a sign that the skin has been damaged. A suntan is perceived as 'healthy' but it may not be so.
4. Long-term sun exposure speeds up the skin's ageing process, making it become more dry and wrinkled.
5. People working outside should consider exposure to UV radiation as an occupational health hazard.

## Who has increased risk of skin damage?

1. People with pale skin, fair hair, freckles or a large number of moles. The risk is less for people with dark hair and brown or black skin. However, prolonged sun exposure can be bad for all skin types, so don't be complacent.
2. People with a family history of skin cancer and those with excessive exposure to sunlight, such as outdoor workers.

## Skin types

**Type 1:** White skin, never tans, always burns: often people with red or fair hair, pale skin and freckles.

**Type 2:** White skin, burns easily, but may tan eventually: people may have fair hair, blue eyes and freckles.

*Types 1 and 2 must take extra care to avoid strong sunshine or cover up.*

**Type 3:** White skin, tans easily and burns rarely: often people with dark hair, eyes and slightly darker skin.

**Type 4:** White skin, never burns, always tans: people with darker hair, eyes and skin.

*Types 3 and 4 should still take care in strong sunshine.*

**Type 5:** Brown skin

**Type 6:** Black skin

*Types 5 and 6 are at little risk of skin cancer but can still darken and even burn in stronger sunlight.*

1. Take care not to burn; this can take as little as 10 minutes.
2. Cover up with loose clothing. Keep your clothing on so that you do not expose unprotected areas.
3. Seek shade during the hottest part of the day and take your breaks in the shade.
4. Frequently apply sunscreen of SPF 15+ to parts of the body exposed to the sun.
5. If you are concerned about moles changing shape or colour and itching, weeping or bleeding, see your GP immediately.

Why do so many people die of skin cancer each year?
What is your skin type, and how should you protect it?

Encourage a discussion by using a real-life situation or example and ask if there are any questions.

# Weil's disease

| | |
|---|---|
| **Reason** | The presence of rats on site must be discouraged. |
| **Why** | Weil's disease can be fatal. Don't become a statistic. |
| **Outline** | This talk covers the effects and symptoms of Weil's disease, the measures you should take to avoid it and who could be at risk. |

*Rats' urine can cause Weil's disease*

| **Notes** |
|---|
| |
| |
| |
| |
| |
| |
| |
| |
| |

# Weil's disease

| | |
|---|---|
| **Reason** | The presence of rats on site must be discouraged. |
| **Why** | Weil's disease can be fatal. Don't become a statistic. |
| **Outline** | This talk covers the effects and symptoms of Weil's disease, the measures you should take to avoid it and who could be at risk. |

## What is Weil's disease?

1. Weil's disease, which is also known as leptospirosis, is a kind of jaundice.
2. The disease enters the body through breaks in the skin and through the lining of the mouth and nose.
3. It is caused by contact with water contaminated by the urine of rats and other small mammals.
4. It starts as a mild illness and it can be easily cured if treated early enough.
5. If left untreated it becomes more serious and can be fatal.
6. The initial symptoms are very similar to flu and so it is possible that you could ignore the symptoms or be treated for the wrong illness.

## What can you do about it?

1. Don't encourage the presence of vermin. Carefully dispose of waste food, especially on sites that are wet or adjacent to rivers and lakes.
2. Do not handle the carcasses of dead rats or other small mammals.
3. If you think you may be at risk, cover all cuts and abrasions with a waterproof dressing and wear appropriate protective clothing.
4. If you frequently work near water, carry a card or tag saying that you may be at risk of catching the disease.
5. Be aware that you can catch the disease if you get water in your mouth and nose after falling in.
6. Symptoms of Weil's disease are mild flu-like symptoms such as headaches, chills and muscle pains.
7. See your doctor immediately if you think you are infected.

## Who is at risk?

1. Anyone who may come into contact with contaminated water.
2. Particularly operatives who work regularly in or near water, such as those engaged in:
   - work on sewers and other drainage systems
   - work on canals and similar conservation projects
   - tunnelling work.

What should you do if you become aware of an increasing rat population?
If you have flu-like symptoms after falling into water that may be contaminated, what should you tell your doctor?

Encourage a discussion by using a real-life situation or example and ask if there are any questions.

# Alcohol and drugs

| | |
|---|---|
| **Reason** | Statistics show that alcohol and drug use are increasing on site. |
| **Why** | This can lead to accidents so we need to make sure it doesn't happen on this site. |
| **Outline** | This talk covers the effects of alcohol and drugs on your safety and that of others. |

**Notes**

# Alcohol and drugs

| | |
|---|---|
| **Reason** | Statistics show that alcohol and drug use are increasing on site. |
| **Why** | This can lead to accidents so we need to make sure it doesn't happen on this site. |
| **Outline** | This talk covers the effects of alcohol and drugs on your safety and that of others. |

## Alcohol

1. In a high risk industry like ours, alcohol and work do not mix well.
2. Alcohol is a drug that depresses parts of the brain function. When working on site you require all of your brain to save you from injury.
3. If you're found to be intoxicated with drink, you won't be allowed on site and you may end up losing your job.
4. Don't get drunk the night before and expect to work safely on site the next day. You could still be over the limit when you drive to work, as alcohol may take more time than you think to work out of your system.
5. Many drivers who are killed in a road accident are over the legal alcohol in blood limit.
6. A pint of 5% beer equals 2.8 units. It takes one hour for the body to get rid of one unit completely.
7. Two pints of strong ale or stout put men over the recommended daily limit. One and a half pints put women over the recommended limit.
8. Some workplace fatal accidents have been alcohol-related.
9. If you get a bad reputation for drinking you may not get another job as you'll be seen as a liability.

## Drugs

1. You are far more likely to have an accident on site when under the influence of drugs.
2. Drugs prescribed by your doctor could make you unfit for work, as can illegal drugs.
3. If you know someone is on drugs don't think that it isn't your problem. Tell your supervisor and help to stamp it out. If you get hurt, it's a bit late to wonder what the other person was on.
4. All drugs can affect your ability to work safely.
5. Some effects of drugs are slow reaction times, clumsiness, poor decision-making and distorted vision.

What effect can alcohol have on you?
What could be the result of being under the influence of alcohol on site?
If you took drugs, what effect could it have on you and your workmates?
How long does a pint of beer take to get out of your system?
What should you do if you see a person taking drugs?

Now inform your workforce of the company policy regarding alcohol and drug misuse.
Encourage a discussion by using a real-life situation or example and ask if there are any questions.

# Needlestick injuries

| | |
|---|---|
| **Reason** | It is possible that you will find a used hypodermic syringe or needle on a site at some time. |
| **Why** | You need to know what to do if you accidentally prick your skin, as you could become infected with a serious disease. |
| **Outline** | This talk covers the actions you should take if you discover a needle and if you prick your skin with it. |

*Treat discarded needles with care*

| Notes |
|---|
| |
| |
| |
| |
| |

If you suffer a needlestick injury and do not follow this guidance, you could be exposed to the HIV virus, Hepatitis B or Hepatitis C – at best very unpleasant, at worst fatal.

# Needlestick injuries

| | |
|---|---|
| **Reason** | It is possible that you will find a used hypodermic syringe or needle on a site at some time. |
| **Why** | You need to know what to do if you accidentally prick your skin, as you could become infected with a serious disease. |
| **Outline** | This talk covers the actions you should take if you discover a needle and if you prick your skin with it. |

## What is a needlestick injury?

An accidental puncture of the skin by a hypodermic needle.

## If you find a needle

1. It has probably been used by a drug user and may be contaminated by infected blood.
2. Do not touch it or move it, unless you have to because of the situation at the time.
3. Leave a responsible person to safeguard it whilst you report the matter to your supervisor.
4. If you have a site nurse, they should be informed.
5. If you do not have a nurse on site, the local Environmental Health Department should be informed.
6. If you must move the syringe or needle:
   - carry it with the needle pointing downwards
   - do **not** wrap it in paper or put it into a litter bin
   - if available, ideally place it in a sharps bin, otherwise place it in a clear glass bottle or jar
   - dispose of it safely through the site nurse, local police or Environmental Health Department
   - wash your hands thoroughly.

## If you prick your skin

1. Do not panic.
2. Gently squeeze the area around the wound to encourage bleeding.
3. Do not suck the wound.
4. Wash the site of the injury thoroughly with soap and water at the first opportunity.
5. Obtain medical assistance as soon as possible from either the site nurse or the nearest hospital with an accident and emergency department. If you can do so safely, take the syringe or needle with you.
6. If dealt with properly and promptly, the risks of a resulting health problem are minimal.
7. Think about the consequences of not acting promptly and possibly being off work for several weeks while you recover.

?

On what types of site do you think you are most likely to discover used needles or syringes?
Due to the secretive nature of drug taking, in which areas of a site do you think that discarded syringes and needles are most likely to be found?
What diseases do you think that you could catch from a needlestick injury?
What could be the worst possible outcome?

Encourage a discussion by using a real-life situation or example and ask if there are any questions.

# General health and wellbeing

| | |
|---|---|
| **Reason** | It is a legal requirement that employers ensure your general health and wellbeing. |
| **Why** | There is a high risk of ill health and injury in the construction industry. |
| **Outline** | This talk covers some of the causes and preventative measures that can be put in place. |

*Make use of any health facilities on site*

| **Notes** |
|---|
| |
| |
| |
| |
| |
| |
| |

# General health and wellbeing

| | |
|---|---|
| **Reason** | It is a legal requirement that employers ensure your general health and wellbeing. |
| **Why** | There is a high risk of ill health and injury in the construction industry. |
| **Outline** | This talk covers some of the causes and preventative measures that can be put in place. |

## General health and wellbeing

1. General ill health is a broad term covering anything from stress, to a bad back or respiratory diseases.
2. Poor work practices are the main cause of ill health, and 500,000 people suffer a new work-related illness each year.
3. People with health problems have to be treated fairly, with consideration by their employer and fellow workers.
4. Your general health is important for the quality of your later life: an injury today can cause many years of suffering, discomfort and sometimes disability.

## Control measures

1. Both you and your employer have a duty to ensure that your general health and wellbeing is provided for in the workplace.
2. Use the correct methods of work and ensure you are provided with the correct protective measures (such as dust suppression, personal protective equipment (PPE), and so on), so that exposure is reduced.
3. Suitable welfare facilities (for example, washing, eating, drinking and changing facilities) should be provided to minimise the risk of ill health.
4. If you take medication or have an existing health problem you must tell your employer so that your wellbeing can be assured.
5. Follow instructions from your supervisor, have a questioning mind and, when in doubt, ask.

## Precautions

1. If you discover a problem (such as a blocked toilet), don't just ignore it – report it.
2. Ensure that the dust mask you have been given is the right one for the job.
3. Check your access and egress routes to make sure they are safe.
4. If you think you need help with a task, ask – don't just try and manage.

?

When should you report something that could cause ill health to you or others?
If you have a problem who should you report it to?
What should you do if you think the welfare facilities are inadequate?
You know that one of your workmates has a family problem causing stress – what should you do?

Now inform your workforce of the company provision for providing good general health and wellbeing on site.
Encourage a discussion by using a real-life situation or example and ask if there are any questions.

# Welfare arrangements, including personal hygiene

| | |
|---|---|
| **Reason** | Adequate welfare facilities are a legal requirement and should be provided on all sites. |
| **Why** | Adequate toilets, washbasins, drying rooms and rest areas are essential for the wellbeing of everyone on site. |
| **Outline** | This talk covers the factors that will determine whether site welfare facilities are adequate. |

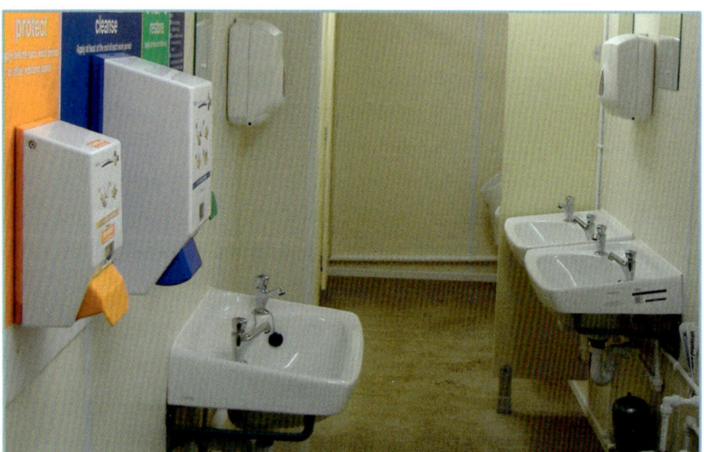


*Examples of good practice welfare facilities*

# Welfare arrangements, including personal hygiene

| | |
|---|---|
| **Reason** | Adequate welfare facilities are a legal requirement and should be provided on all sites. |
| **Why** | Adequate toilets, washbasins, drying rooms and rest areas are essential for the wellbeing of everyone on site. |
| **Outline** | This talk covers the factors that will determine whether site welfare facilities are adequate. |

## General welfare

1. There must be an adequate number of toilets, washbasins, drying spaces and rest areas provided, in relation to the number of people on site. Showers may have to be provided on some jobs.
2. Toilets and washbasins must be properly maintained and kept clean. Hot (or warm) and cold running water must be provided.
3. Soap and a way of drying your hands must be provided.
4. If you must change out of 'street' clothes, a changing area must be provided with storage for your 'street' clothes.
5. Rest areas should include one or more rest rooms.
6. Smoking is now not allowed in enclosed welfare areas, such as canteens.

## Employees' responsibilities

1. Welfare facilities should not be dirty, vandalised or covered by graffiti – leave them as you wish to find them.
2. Only smoke in authorised areas.
3. Tell your supervisor if you are aware that welfare facilities are being deliberately damaged or otherwise misused.

## Food safety

1. Larger sites may provide hot or cold food as part of welfare arrangements.
2. All food must be stored, handled and prepared in hygienic conditions.
3. Anyone preparing food for others must be trained, observe good standards of personal hygiene, not smoke in food preparation areas and report certain illnesses.
4. Where a cooker or microwave oven is provided for you to prepare your own food, ensure the food is thoroughly cooked – undercooked food can lead to food poisoning.
5. Dispose of waste food safely. Do not encourage rats or other vermin.

What changing areas are available for your use on this site?
What are your responsibilities with respect to welfare facilities?
What should you do if you see a toilet block being vandalised?
How can rats be a danger to your health?

Encourage a discussion by using a real-life situation or example and ask if there are any questions.

# First aid

| | |
|---|---|
| **Reason** | First aid is an important factor on construction sites. |
| **Why** | If you know basic first aid, you could save a life. |
| **Outline** | This talk explains what you should know, basic first aid and when it is required. |

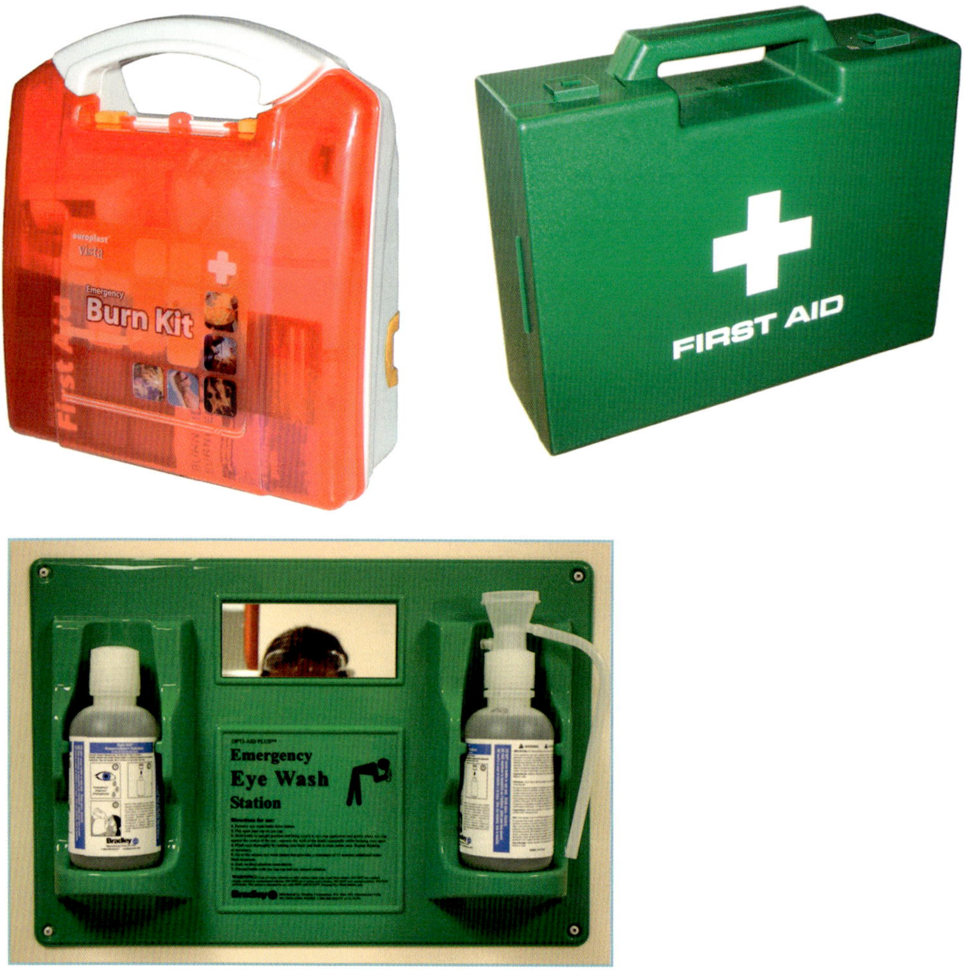


*Types of first-aid kit*

If you know first aid you could save a life.

# First aid

| | |
|---|---|
| **Reason** | First aid is an important factor on construction sites. |
| **Why** | If you know basic first aid, you could save a life. |
| **Outline** | This talk explains what you should know, basic first aid and when it is required. |

## Before first aid is required

1. Ensure you know where the first-aid kit is kept.
2. Know who the first aider and appointed persons are.
3. Have a small, travelling first-aid kit if you are working in a small group away from the main site or if you use potentially dangerous tools or machinery.
4. Familiarise yourself with the procedure for calling the emergency services.

## Discovering a casualty

1. Call for medical help and ensure your own safety as you approach the casualty.
2. Remove the hazard from the casualty, if it is safe to do so.
3. Send someone to phone for an ambulance if necessary.
4. Don't move the casualty, unless they are in immediate danger.
5. Remain with the casualty, give reassurance and make them as comfortable as possible.
6. Don't give drinks or food to the casualty. Only moisten their lips and don't allow them to smoke.

## Your employer is expected to have:

1. completed a first-aid needs assessment
2. ensured that there is either an appointed person to take charge of first-aid arrangements, or that there are appropriate numbers of suitably trained first aiders
3. ensured that there are adequate facilities and a suitably stocked first-aid box
4. provided you with information about the first-aid arrangements.

## Employers' responsibilities

1. Ensure that first-aid cover includes, for example, shift work and holiday cover.
2. Appoint somebody to restock the first-aid boxes.
3. Ensure the first-aid equipment provided is appropriate for the nature of the work and the number of workers.
4. Ensure first aiders are easy to identify – usually by a sticker on their safety helmet.

When would you need a first-aid kit of your own?
On finding a casualty, what is the first thing to do?
What should you do until a first aider arrives?
What is your employer expected to provide you with?
What are the priorities of first aid?

Now inform your workforce of the company provision for first aiders and first-aid kits and actions to be taken in the event of an emergency, including the procedure for calling the emergency services.
Encourage a discussion by using a real-life situation or example and ask if there are any questions.

# Personal protective equipment

| | |
|---|---|
| **Reason** | Some dangers arising from hazardous activities can only be controlled by the use or wearing of personal protective equipment (PPE). |
| **Why** | In many cases, it is not possible to completely eliminate hazards by other means. |
| **Outline** | This talk covers the measures necessary to ensure that the use of PPE is effective. |

*PPE should be appropriate to the job*

*Simple PPE can prevent long-term damage*

PPE can be a life-saver: use it properly and look after it, and it should look after you.

# Personal protective equipment

| | |
|---|---|
| **Reason** | Some dangers arising from hazardous activities can only be controlled by the use or wearing of personal protective equipment (PPE). |
| **Why** | In many cases, it is not possible to completely eliminate hazards by other means. |
| **Outline** | This talk covers the measures necessary to ensure that the use of PPE is effective. |

## What is PPE?

1. PPE is equipment or clothing worn to protect the user from known hazards in the workplace.
2. In construction, the most commonly worn items of PPE are safety helmets, high-visibility clothing and safety footwear.
3. Other examples of PPE are respirators, safety harnesses, earplugs, safety glasses and goggles, protective gloves and also some clothing.

## Limitations

1. PPE will only protect the user.
2. It must be used in accordance with the manufacturer's instructions and any training provided.
3. The actual level of effectiveness is difficult to assess.
4. It must be in good condition to be fully effective; do not mistreat PPE, your health or life might depend upon it.

## What your employer must do

1. Assess the risks to your health and safety arising from your work activities.
2. Try to organise work activities so that PPE is not necessary.
3. Where PPE is necessary, select appropriate items that suit the wearer and are CE-marked.
4. Supply you with the necessary PPE free of charge.
5. Train you to use relevant PPE, explain its limitations and explain the implications of not using it.
6. Ensure compatibility if more than one item of PPE is worn.
7. Ensure that PPE is maintained and replace defective or lost PPE, at no cost to you.

## What you must do

1. Use PPE in accordance with instructions and training given.
2. Return PPE to its storage, where provided, after use.
3. Take reasonable care of your PPE and report its loss or any damage to it to your employer.
4. Never work without PPE when it is known to be necessary.

How much can your employer charge you for PPE?
What factors will determine the life of a safety helmet?
How do you think wearing PPE could increase the dangers to you?
What should you do if you find an item of your PPE is damaged?
What should you look for on any item of PPE to confirm that it has been made to the required standard?

Encourage a discussion by using a real-life situation or example and ask if there are any questions.

# Eye protection

| | |
|---|---|
| **Reason** | On average 1,000 injuries to people's eyes occur every day; 75% by impact, 10% by ingress of foreign bodies (dust) and 15% by burns or chemicals. |
| **Why** | The majority of these injuries could have been prevented if eye protection had been worn. |
| **Outline** | This talk covers the hazards and the reasons for wearing eye protection. |

*Always wear appropriate eye protection*

*Safety glasses cleaning station*

Eye protectors are replaceable, your eyes are not.

# Eye protection

| | |
|---|---|
| **Reason** | On average 1,000 injuries to people's eyes occur every day; 75% by impact, 10% by ingress of foreign bodies (dust) and 15% by burns or chemicals. |
| **Why** | The majority of these injuries could have been prevented if eye protection had been worn. |
| **Outline** | This talk covers the hazards and the reasons for wearing eye protection. |

## Possible causes of eye injury

1. Using drills, hammers, chisels or other percussive tools.
2. Using compressed air and any gas, liquid or vapour under pressure.
3. Working with power tools where debris may be given off at speed (for example, using a grinder).
4. Working with tools that will result in chippings being broken off (for example, using breakers).
5. Welding, when ultraviolet light is given off that can damage your eyes.
6. Handling, or coming into contact with, corrosive or irritant substances (such as acids or alkalis).
7. Handling and using cartridge-operated or compressed gas tools (such as a nail gun).
8. Cutting of wire or metal strapping that is under tension (such as brick bands).

## Wearing eye protection

1. You have a legal obligation to use any eye protection provided.
2. Don't go into areas where eye protection is required unless you are wearing it.
3. Ensure your eye protection fits you and is suitable for the job.
4. Ensure that any lost or damaged eye-protection is replaced immediately; don't be tempted to work without it.
5. Your eye protection should be appropriate for the job, comfortable to wear and kept clean.
6. There are several grades of eye protection, each to combat a different type of hazard. Only use the eye protection designated for the job you are doing.
7. All eye protection must carry a CE-mark to show it has been made to an industrial standard.
8. Take care of your eye protection when it is not being worn and use a protective case.
9. Light eye protection should not be used instead of impact goggles.

?

What are the hazards associated with compressed air?
How can welding affect unprotected eyes?
What must you ensure on being issued eye protection?
If you get something in your eye, what should you do?
What should you do if you damage your eye protectors?
What are the hazards when using a cut-off wheel?

Encourage a discussion by using a real-life situation or example and ask if there are any questions.

# Respiratory protection

| | |
|---|---|
| **Reason** | Inhalation of harmful dusts and fumes causes respiratory problems, which must be prevented. |
| **Why** | Breathing in harmful substances causes health problems and long-term suffering. |
| **Outline** | This talk covers aspects that require consideration when protecting against respiratory risks. |

*Respiratory protective equipment (RPE) and specialist tools can prevent respiratory problems*

| Notes |
|---|
| |
| |
| |
| |
| |
| |
| |
| |
| |
| |

A mask protects the user, but safe systems protect everyone.

# Respiratory protection

| | |
|---|---|
| **Reason** | Inhalation of harmful dusts and fumes causes respiratory problems, which must be prevented. |
| **Why** | Breathing in harmful substances causes health problems and long-term suffering. |
| **Outline** | This talk covers aspects that require consideration when protecting against respiratory risks. |

## Some facts

1. Every year up to 7,000 people develop occupational asthma from breathing in dust and fumes.
2. 4,500 people die every year from asbestos-related diseases from breathing in airborne asbestos fibres.
3. About 27 million working days a year are lost to work-related ill health (across all industries).
4. At any one time there are far more people off work through occupational ill health than there are because of work-related accidents.

## Provision and use

1. Where dust and fumes cannot be avoided by engineering or mechanical means (such as extraction, collection or dampening down) then the correct respiratory protective equipment (RPE) must be provided and used.
2. RPE must be designed to reduce the level of exposure to the type and quantity of dust or fumes.
3. Different types of RPE are available, ranging from disposable, filtering half-masks and half-mask respirators with cartridges, to ventilated visors or helmet respirators.
4. Selection of RPE is important and must take into account people who wear glasses, have stubble or a beard, or have long hair when it is not swept back and secured.
5. Personal face fit is a critical element in the selection and use of RPE.

## Your duties

1. Use RPE properly, whenever it is required.
2. Report any defects in, or damage to, the RPE immediately.
3. Participate in any training or instruction provided on RPE.
4. Inform your employer of any medical conditions that you have, which might be affected by the use of RPE.

What should you think about, for yourself and others, in your work area if you are asked to carry out a task that involves creating dust?
What should you do if your face mask does not fit comfortably?
If you have prescription safety glasses and you have recently started wearing glasses what should you do?

Now inform your workforce of the company provision and policy for RPE assessment, face-fit training and issue.
Encourage a discussion by using a real-life situation or example and ask if there are any questions.

# Dust and fumes

| | |
|---|---|
| **Reason** | Exposure to dust and fumes must be prevented or controlled. |
| **Why** | Breathing in dust or fumes can lead to long-term health problems (such as asthma). |
| **Outline** | This talk covers some sources and dangers from dust and fumes, and examples of precautions that can be taken. |

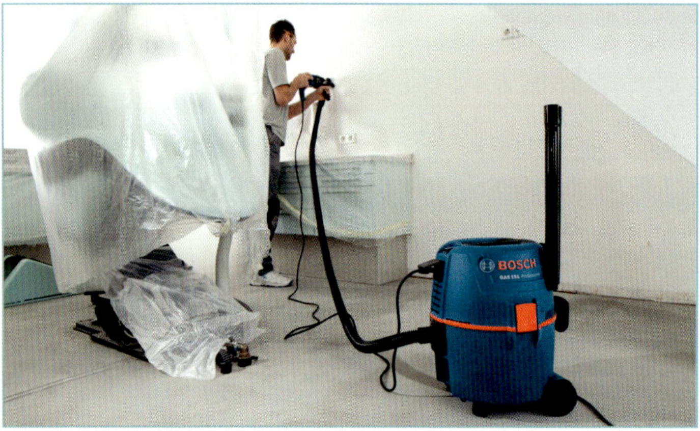


*Drill with dust extraction*

*Insulation dust*

**!** You can leave a dusty place at any time, but asthma lasts forever.

# Dust and fumes

| | |
|---|---|
| **Reason** | Exposure to dust and fumes must be prevented or controlled. |
| **Why** | Breathing in dust or fumes can lead to long-term health problems (such as asthma). |
| **Outline** | This talk covers some sources and dangers from dust and fumes, and examples of precautions that can be taken. |

## Some sources of harmful dust and fumes

1. Cutting, sanding and grinding of some materials will create harmful dust.
2. Welding and gas cutting of metals can create harmful fumes.
3. Work with old lead can expose you to lead oxide dust (white, powdery deposits) which is also harmful.
4. Burning off old lead-based paints or heating lead can create harmful fumes.
5. Stripping out or other work involving fibrous insulation (such as asbestos or fibreglass insulation) can release harmful dust into the air.

## Health risks from breathing in dust or fumes

1. Silica dust from cutting or scabbling concrete or cutting bricks or stone can cause lung disease (for example, asthma).
2. Dust from cutting or sanding hardwood can cause nasal cancer.
3. Asbestos dust can cause cancer of the lungs or lining of the chest cavity.
4. Welding fumes can result in metal fume fever, which has flu-like symptoms.
5. Breathing in the fumes from solvents and paint can lead to nausea, drowsiness, headaches and, eventually, unconsciousness and death in extreme cases.
6. Investigations are continuing into the possible harmful effects of breathing in dust from synthetic insulation materials (such as fibreglass matting) – regard it as hazardous.

## Precautions

1. Where it is possible, the job should be planned to eliminate harmful dust and fumes.
2. If elimination is not possible, harmful dust and fumes must be controlled so they are not breathed in.
3. Some tools and plant are fitted with dust extraction and collection devices. If these are available, use them.
4. If your employer has provided portable extraction equipment, use it.
5. It may be necessary for you to wear RPE to protect yourself from the effects of dust or fumes. Make sure you know how to use it properly.
6. Consider the effects that your work may have on other people.

What work do you carry out that creates harmful dust or fumes?
What types of RPE are suitable for use with hazardous dust and fumes?
How do you inform others that you will be creating harmful dust or fumes?

Encourage a discussion by using a real-life situation or example and ask if there are any questions.

# Asbestos

| | |
|---|---|
| **Reason** | Up to 4,500 people a year, from all industries, die from asbestos-related diseases. |
| **Why** | You need to be aware of asbestos – it could kill you. |
| **Outline** | This talk covers where you will find asbestos, how it can affect you and hazardous work. |

*Asbestos fire blanket*

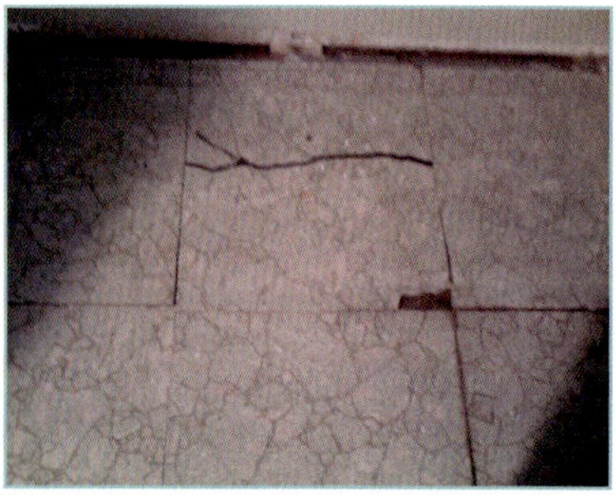

*Asbestos floor tiles*

*Asbestos soffit*

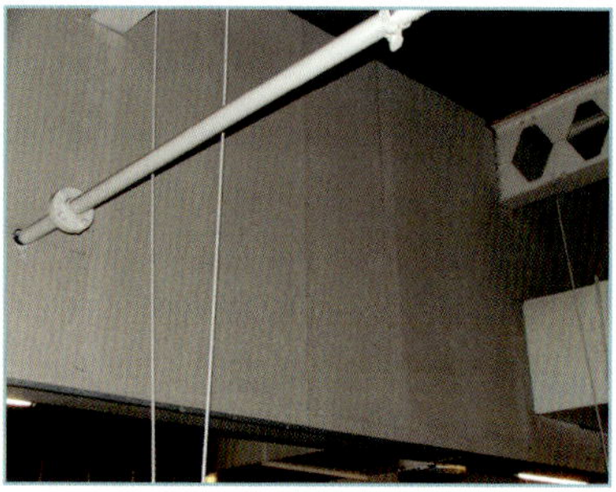

*Asbestos (AIB) panels at high level*

*Damaged asbestos spray applied acoustical ceiling material*

**!** If you suspect asbestos, stop work immediately and tell your supervisor.

# Asbestos

| | |
|---|---|
| **Reason** | Up to 4,500 people a year, from all industries, die from asbestos-related diseases. |
| **Why** | You need to be aware of asbestos – it could kill you. |
| **Outline** | This talk covers where you will find asbestos, how it can affect you and hazardous work. |

## Where you will find asbestos

1. In insulation and sprayed coatings used for:
   - boilers, plant and pipework hidden in underfloor ducting
   - fire protection to steelwork, hidden behind false ceilings
   - thermal and acoustic insulation of buildings
   - some textured coatings and paints
   - friction materials (such as brake linings and clutch plates)
   - gaskets and packings in engines, heating and ventilation systems.
2. In insulating board used in:
   - fire protection to doors, protected exits and steelwork
   - claddings on walls and ceilings
   - internal walls, partitions and suspended ceiling tiles.
3. In asbestos cement, which is found as:
   - corrugated roofing and cladding sheets of buildings
   - flat sheets for partitions, cladding and door facings
   - rainwater gutters and downpipes.

## How asbestos can affect you

1. Asbestos breaks into tiny, long, sharp fibres. They can get lodged and scar the lungs, causing asbestosis or fibrosis.
2. Asbestos fibres may cause a form of lung cancer.
3. It can also cause mesothelioma, a cancer of the inner lining of the chest wall. This cancer is incurable.
4. Smoking increases the risk of asbestos-related diseases.

## Hazardous work

1. Anyone working on building repair and refurbishment is considered most at risk.
2. Buildings constructed up until 1999 may have forms of asbestos-containing materials in them.
3. The removal of roofing felts, old floor tiles, textured paints and plasters containing asbestos can be hazardous.
4. If you think you've come across asbestos, stop work and tell your supervisor.
5. Beware of asbestos-containing materials that have been painted using lead-based paint. This could be a double hazard (*refer to Toolbox talk B18*).

?

Where will you find insulation, sprayed coatings, insulating boards or asbestos cement?
In what ways can asbestos fibres affect you?
What are some of the hazardous work areas?
What should you do if you think that you have discovered asbestos materials in your work area?
What diseases can exposure to asbestos cause?

Now inform your workforce of the company policy regarding the discovery of suspected asbestos. Encourage a discussion by using a real-life situation or example and ask if there are any questions.

# Silica dust

| | |
|---|---|
| **Reason** | Breathing in silica dust must be prevented or controlled. |
| **Why** | Inhalation of silica dust can lead to long-term health problems. |
| **Outline** | This talk covers sources of silica dust, associated health risks and how to control exposure. |

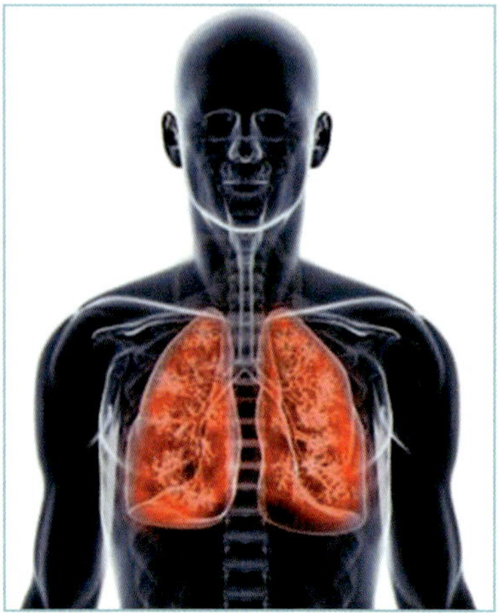

*Silica dust can cause lung damage*

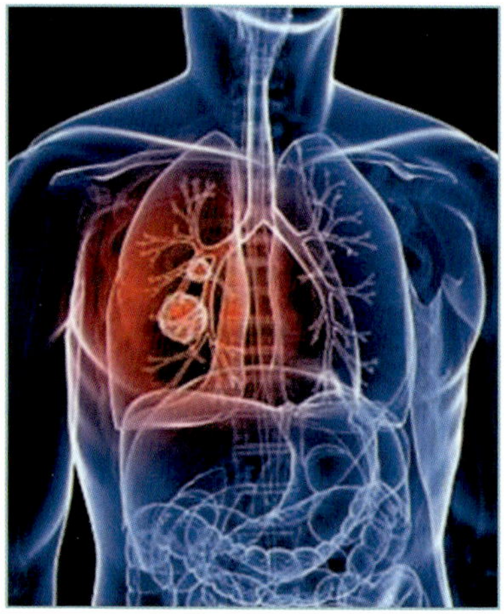

*Damage to lung tissue*

*Wet cutting is a good control measure*

*Wear RPE provided*

Smoking causes lung damage and adds to the damage caused by breathing in silica dust.

# Silica dust

| | |
|---|---|
| **Reason** | Breathing in silica dust must be prevented or controlled. |
| **Why** | Inhalation of silica dust can lead to long-term health problems. |
| **Outline** | This talk covers sources of silica dust, associated health risks and how to control exposure. |

## What is silica?

1. Crystalline silica is a basic component of soil, clay, sand, shale, slate, granite and many other minerals, including components used to make concrete and mortar.
2. Quartz is the most common form.
3. Many materials in the construction industry contain crystalline silica, including bricks and concrete blocks.
4. When workers chip, cut, drill, grind, grit blast, scabble or tunnel through objects that contain crystalline silica the particles can become small enough to breathe in.
5. The use of power tools can lead to high exposure if exhaust systems or wet-cutting processes are not used or maintained.

## Health hazards and symptoms

1. Crystalline silica has been classified as a **human lung carcinogen**.
2. Breathing crystalline silica dust can also cause **silicosis**, which, in severe cases, can be disabling or even fatal.
3. When silica dust enters the lungs it causes scar tissue, reducing the ability of the lungs to take in oxygen.
4. There is no cure for silicosis. Since silicosis affects lung function, it makes a person more susceptible to lung infections like **tuberculosis**.
5. In addition, smoking causes lung damage and adds to the damage caused by breathing in silica dust.

## Preventative and protective measures

1. Replace crystalline silica materials with safer substitutes whenever possible.
2. Use engineering or administrative controls (such as local exhaust ventilation and wet cutting).
3. Use RPE to reduce exposures to a safe level, where necessary.
4. Wear disposable or washable work clothes and use shower facilities, if they are available.
5. Participate in training, exposure monitoring, and health screening and surveillance programmes to monitor any adverse health effects caused by exposure.
6. Be aware of the tasks creating crystalline silica dust and consider who may be affected, including the general public.
7. Do not eat, drink or smoke in areas where crystalline silica dust is present. Wash your hands and face outside dusty areas before performing any of these activities.

If you are asked to mechanically cut a kerbstone with a petrol-powered saw, what precautions should you take to protect yourself?
Why is it important to consider members of the public when creating dust by cutting with a power saw?
What PPE and RPE should you wear when cutting or chasing concrete?
Why is it important not to eat, drink or smoke in dusty areas?

Now inform your workforce of the company policy on the provision of RPE.
Encourage a discussion by using a real-life situation or example and ask if there are any questions.

# Noise

| | |
|---|---|
| **Reason** | Noise-induced hearing loss is a common occupational health hazard. |
| **Why** | There is no satisfactory treatment for noise-induced hearing loss. When you're deaf, you stay deaf. |
| **Outline** | This talk covers the hazards, controlling noise and ear protection. |

*Excessive wear – loose fitting head band*

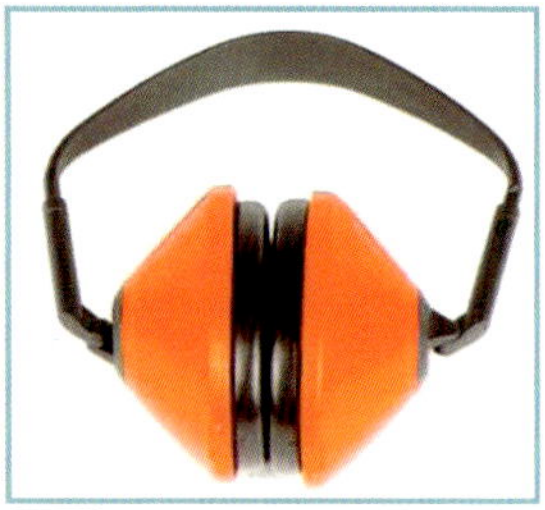

*Tight fitting head band*

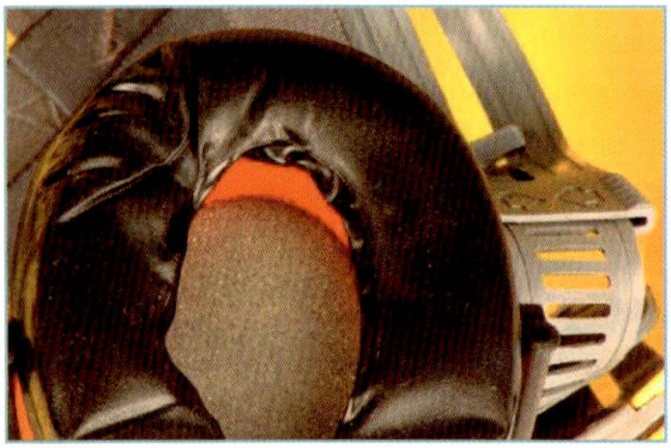

*Damaged seal on ear defenders*

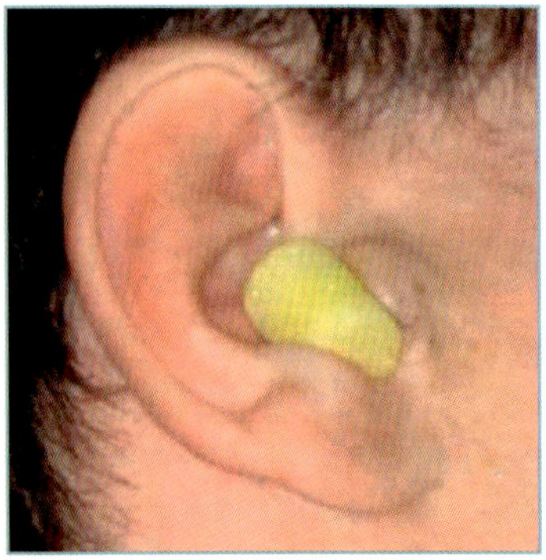

*Earplugs fitted incorrectly*

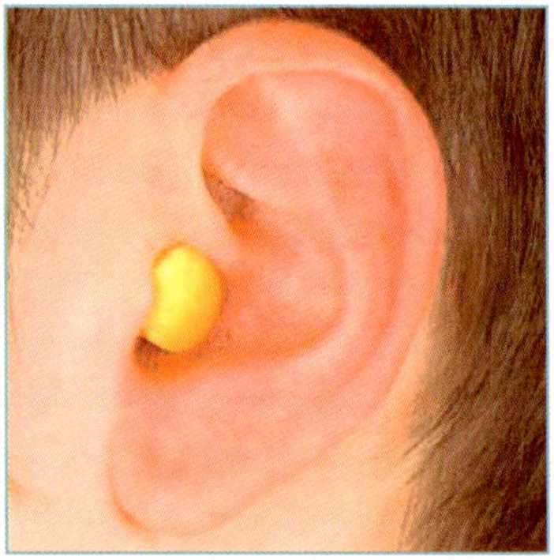

*Earplugs fitted correctly*

Protect your hearing. Loss of hearing lasts forever.

# Noise

| | |
|---|---|
| **Reason** | Noise-induced hearing loss is a common occupational health hazard. |
| **Why** | There is no satisfactory treatment for noise-induced hearing loss. When you're deaf, you stay deaf. |
| **Outline** | This talk covers the hazards, controlling noise and ear protection. |

## Hazards

1. Compressors, breakers, circular saws, generators, vibrating rollers and excavators, angle grinders and power saws can all be harmful to your hearing.
2. Even if you are not using the noisy piece of equipment, you could be affected by someone using it close by.
3. Look out for noise hazard signs on site and obey them.
4. Remember to protect your hearing after work as well (for example, in noisy clubs or when using personal music players).
5. A ringing in the ears after being exposed to noise is an early sign of hearing damage.

## Controlling noise

1. If shouting is necessary in order to be heard from one metre away, the noise level is high and you should be wearing ear protectors.
2. Keep compressor covers closed when in use and ensure breaker mufflers are correctly fitted.
3. Don't leave machinery running unnecessarily and try not to expose others to your noise.
4. If possible, move the noise source away from the work area or move the work area away from the noise.
5. If possible, shield noisy processes; work behind sound-absorbing materials (such as spoil heaps).

## Ear protection

1. Generally, earplugs or muff-type hearing protectors will be issued.
2. Ensure earplugs are a good fit and correctly inserted by following the pack instructions.
3. Regularly clean reusable earplugs.
4. Use disposable earplugs for one shift only.
5. Clean your hands before touching all types of earplugs.
6. Hearing protectors should fit the head all round the seal.
7. Adjust the head or neck band as necessary and wear it in the correct position.
8. Ensure hearing protector seals are in good condition; remove and wash them in soapy water regularly.
9. Don't alter the pressure of the cups on the ear by bending the headband.
10. If you have difficulty in wearing hearing protectors, report it.

What sources of noise can be found on site?
When should ear protection be worn?
Name two precautions you can take with machinery to reduce noise.
What should not be used instead of earplugs?
What must you ensure when wearing ear defenders?
How can you reduce noise levels from a machine?
What must you remember when handling earplugs?

Encourage a discussion by using a real-life situation or example and ask if there are any questions.

# Hand-arm vibration

| | |
|---|---|
| **Reason** | Exposure to vibration can result in serious and disabling injury. |
| **Why** | Many workers do not appreciate the possible dangers from vibration. |
| **Outline** | This talk covers the effects, sources and methods of overcoming excessive vibration. |

*Be aware of the risks of vibrating tools*

The long-term effects of exposure to vibration can be permanent and disabling.
Don't let it happen to you.

# Hand-arm vibration

| | |
|---|---|
| **Reason** | Exposure to vibration can result in serious and disabling injury. |
| **Why** | Many workers do not appreciate the possible dangers from vibration. |
| **Outline** | This talk covers the effects, sources and methods of overcoming excessive vibration. |

## Effects of vibration

1. Depending upon the work situation, vibration can be whole-body vibration or, more commonly, hand-arm vibration (HAV).
2. The first signs of a HAV problem may only be tingling in the affected fingers.
3. Exposure to vibration can lead to irritation, fatigue and loss of concentration.
4. These symptoms are likely to affect a person's attention to safety.
5. In the longer term, HAV damage may occur to blood vessels, nerves, muscles, tendons and body organs.
6. Excessive exposure to HAV can lead to vibration white finger, resulting in damaged blood vessels, circulatory problems, pain and possibly gangrene.

## Sources of vibration

1. A common cause of HAV is the prolonged use of rotating hand tools used for cutting and grinding.
2. Percussive hand tools used for riveting, chipping, hammering and drilling are also sources of HAV.
3. The use of chainsaws is another source of HAV.

## How to avoid vibration

1. Advances in technology are leading to newer tools being equipped or manufactured with vibration-absorbing features. If available, select tools with vibration-absorbing features for your work.
2. When using a tool that causes vibration, break the job up with other work activities.
3. Adopt a comfortable stance when using vibrating tools – tense muscles increase ill effects.
4. Keeping warm and dry also helps, especially for the hands and arms.
5. If you think you are suffering ill effects from vibration, stop the activity, speak to your supervisor and if necessary seek medical advice.
6. Anti-vibration gloves do not work and should not be relied upon as a control measure, although they do help to keep hands warm.

During your work when might you be subjected to vibration?
What tools do you use that cause hand-arm vibration?
What should you do if you notice that your fingers are tingling after you have finished a long job where you have used a power tool?
How can excessive vibration be avoided?

Encourage a discussion by using a real-life situation or example and ask if there are any questions.

# COSHH

| | |
|---|---|
| **Reason** | Hazardous substances can be used in, or created by, construction processes. |
| **Why** | Using hazardous substances can damage your health. |
| **Outline** | This talk covers risk assessment, hazards, control measures and safe use of substances. |

| New globally harmonised symbols | | Existing European hazard (CHIP) symbols | |
|---|---|---|---|
|  | Fatal if inhaled, swallowed or in contact with skin. |  | **Very toxic or toxic**<br>Substances that, in very low quantities or low quantities, cause death or acute or chronic damage to health when inhaled, swallowed or absorbed via the skin. |
|  | Warning – harmful if inhaled, swallowed or in contact with skin. Can also cause serious eye irritation. |  | **Harmful**<br>Substances that may cause death or acute or chronic damage to health when inhaled, swallowed or absorbed via the skin. |
|  | Warning – harmful if inhaled, swallowed or in contact with skin. Can also cause serious eye irritation. |  | **Irritant**<br>Non-corrosive substances that may cause inflammation through immediate, prolonged or repeated contact with the skin or mucous membrane. |
|  | Danger – causes severe skin burns and eye damage. |  | **Corrosive**<br>Substances that may, on contact with living tissues, destroy them. |
|  | Danger (Category 1) – may cause allergy or asthma symptoms or breathing difficulties if inhaled.<br>Danger (Category 1a and 1b) – may cause cancer. | None | |

Know what PPE to wear to protect against the hazard.

# COSHH

| | |
|---|---|
| **Reason** | Hazardous substances can be used in, or created by, construction processes. |
| **Why** | Using hazardous substances can damage your health. |
| **Outline** | This talk covers risk assessment, hazards, control measures and safe use of substances. |

## Risk assessment

1. A competent person must carry out a risk assessment to decide whether:
   - exposure to a substance can be avoided
   - alternative work methods can reduce exposure
   - a less hazardous substance can be used instead.
2. Any substance with a hazard warning label has the potential to cause harm.

## Hazards

1. Your health could be affected by a hazardous substance through:
   - ingestion – eating contaminated food
   - inhalation – breathing in harmful dust or fumes
   - absorption – chemicals entering through cuts.
2. Examples of hazardous substances on construction sites include:
   - contaminated ground
   - cement
   - hardwood dust
   - concrete admixtures
   - solvent fumes
   - resins
   - epoxy-based paints
   - welding fumes
   - asbestos.
3. Don't mix chemicals or substances unless it is a safe and authorised process.

## Control measures

1. Follow your employer's safe system of work.
2. When using hazardous substances, wear the correct PPE.
3. Know how to look after and use PPE correctly.
4. Ensure hazardous substances are put back into a secure location after use. Do not leave them on site.

## Safe use of substances

1. Make sure you are trained to use hazardous substances.
2. Comply with the Control of Substances Hazardous to Health (COSHH) assessment and the instructions on the product label.
3. Don't eat, drink or smoke when handling substances.
4. Don't expose others to fumes, dust, gas or other dangers from hazardous substances due to your work.
5. Always wash at the end of each shift and before eating.

Before using a substance, what should you consider?
Name the three ways a substance can enter your body.
Where should substances be put at the end of a shift?
Where can you obtain information from about the hazardous substance you are using?
What can you wear to protect yourself against substances?
Name five hazardous substances you may find on site.

Encourage a discussion by using a real-life situation or example and ask if there are any questions.

# Lead, lead paint and lead dust

| | |
|---|---|
| **Reason** | Anyone working with lead or lead-containing materials must understand the dangers. |
| **Why** | Lead is a cumulative poison and so exposure must be controlled. In extreme cases it can kill. Workers, building occupants and their families can be at risk. |
| **Outline** | This talk covers the effects of lead on the body, the sources of exposure and how to control exposure. |

| **Notes** |
|---|
| |
| |
| |
| |
| |
| |
| |
| |
| |
| |
| |
| |
| |
| |
| |
| |
| |

Why is it important to control exposure to lead?
Who has responsibility for your health and safety?
What work do you do that may expose you to lead?
What is the most likely route of lead entry into the body resulting from your work?

Encourage a discussion by using a real-life situation or example and ask if there are any questions.

# Lead, lead paint and lead dust

| | |
|---|---|
| **Reason** | Anyone working with lead or lead-containing materials must understand the dangers. |
| **Why** | Lead is a cumulative poison and so exposure must be controlled. In extreme cases it can kill. Workers, building occupants and their families can be at risk. |
| **Outline** | This talk covers the effects of lead on the body, the sources of exposure and how to control exposure. |

## The effects of lead

1. Lead has long been known to be a poison (toxic).
2. Exposure can cause headaches, tiredness, irritability and nausea. More serious effects are damage to the kidneys, nerves and brain.
3. Ingesting lead may cause impotence and male infertility and cause problems in pregnancy for females. Females of childbearing age must be particularly protected from exposure.
4. Smoking increases the risk of lead-related health problems.

## Sources of lead exposure

1 Most risks are from dust, fumes or vapour, from when lead is heated, cut, damaged or otherwise disturbed.
2. As a result of hot works (such as torch-cutting, burning or welding).
3. Through repair, refurbishment, decoration and demolition work involving old lead or lead-painted structures or features. There is an increased risk within pre-1970 buildings and structures.
4. From dry-sanding (for example windows and doors in private homes). Building occupiers can be at risk if clean-up is not thorough.
5. Cutting lead or lead-painted materials with disc-cutters.
6. Burning off old lead-based paints (which should be avoided).
7. Spray painting with lead-based paints.
8. Lead-painted asbestos-containing materials – double hazard.
9. Disturbance of leaded exhaust particles in roof insulation.
10. Heating lead to lower temperatures (such as plumbing and soldering) and work involving handling clean sheet lead are regarded as lower risk activities, but still may require control measures to be put in place.

## Control of lead exposure

1. Employers must prevent or control lead exposure for anyone likely to be directly or indirectly affected.
2. Any lead dust must be thoroughly cleaned up and lead fumes controlled.
3. The routes of entry into the body are inhalation (through the nose) and ingestion (through the mouth).
4. Don't take lead dust home on your clothes, as this can cause a risk to your family.
5. Your employer must inform you of health risks and advise lead-safe working practices, such as working 'wet', HEPA vacuuming and using LEV that has been provided.
6. You may have to wear RPE (minimum FFP3 or P3) to protect against lead dust, depending on air sampling results.
7. After working with lead, wash contaminated skin before eating or drinking. Good personal hygiene is essential.
8. Never eat, drink or smoke in areas where work with lead has been carried out.
9. If you are at significant exposure risk your blood-lead levels will need to be monitored to determine your exposure.

# Manual handling

| | |
|---|---|
| **Reason** | Every year, a significant proportion of all injuries at work are caused by manual handling. |
| **Why** | If you get it wrong today you will suffer the consequences tomorrow. |
| **Outline** | This talk covers considerations and good techniques for manual handling. |

*Simple aids like wheelbarrows reduce the risk of injury*

*Use mechanical aids where possible*

*Some jobs can't be done safely on your own*

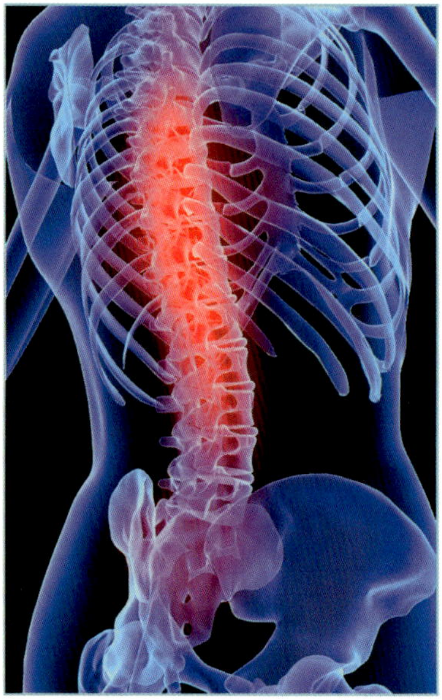

*Back injuries can be permanent*

Poor manual handling techniques can cause serious injuries.

# Manual handling

| | |
|---|---|
| **Reason** | Every year, a significant proportion of all injuries at work are caused by manual handling. |
| **Why** | If you get it wrong today you will suffer the consequences tomorrow. |
| **Outline** | This talk covers considerations and good techniques for manual handling. |

## Considerations

1. Always use mechanical aids instead of manual handling if possible (such as forklifts or pallet trucks).
2. Know your capabilities and only tackle jobs you can handle, otherwise seek help.
3. Check if there is a clear walkway with good lighting to the work area.
4. Where possible, establish the weight of the load before starting to lift.
5. Wear gloves to protect against cuts and safety boots to protect from falling loads.
6. Carry out a trial lift by rocking the load from side to side, then try lifting it a small way to get a feel for it.

## Good handling techniques

1. Stand reasonably close to the load, feet hip-width apart, one foot slightly forward pointing in the direction you're going.
2. Flex your knees and slightly curve your back (avoid stooping or a deep squat when starting a lift).
3. Get a secure grip on the load.
4. A good lifting technique uses the strong muscles in your legs.
5. Keep the load close to your body but don't carry a load that obscures your vision.
6. Lift slowly and smoothly, avoiding jerky movements.
7. Avoid twisting your body when lifting or carrying a load.
8. When lifting to a height from the floor, do it in two stages if possible.
9. When two or more people lift a load, one person must give directions to co-ordinate the lift.

What checks should you carry out before lifting?
What PPE should you wear and why?
Describe how you would lift an object safely.
When two or more people are lifting a load, what should happen?

Now inform your workforce of the company policy regarding manual handling.
Encourage a discussion by using a real-life situation or example and ask if there are any questions.

# C
# General safety

# Site set up and security

| | |
|---|---|
| **Reason** | Site security is essential for the protection of people and materials. |
| **Why** | Unauthorised persons will probably not be aware of the hazards associated with construction sites. |
| **Outline** | This talk covers the hazards to unauthorised visitors and ways of preventing unauthorised access. |

*Children have been injured and killed after gaining access to construction sites*

*Safe egress (exit) is important*

*Security and safety signs in place*

No job is so urgent or so important that it cannot be done safely.

# Site set up and security

| | |
|---|---|
| **Reason** | Site security is essential for the protection of people and materials. |
| **Why** | Unauthorised persons will probably not be aware of the hazards associated with construction sites. |
| **Outline** | This talk covers the hazards to unauthorised visitors and ways of preventing unauthorised access. |

## What the law says

1. Under the law trespassers (particularly children who are less aware of danger) have a right not to be put at risk if they enter a construction site.
2. The CDM Regulations place a specific duty on the principal or main contractor to ensure that unauthorised persons do not gain access to the site.

## Some risks to the unwary

1. Children often find that construction sites are exciting places to play. You must ensure they cannot gain access during or after normal working hours.
2. Power tools, plant and equipment may be too tempting if not disabled or locked away; they could be stolen or cause injury to the inexperienced.
3. Hazardous substances that you may be familiar with and use daily may cause serious injury to unauthorised persons; lock them away when not in use.
4. Ensure that bottled gas compounds are kept securely locked.
5. Ensure that all site accommodation is locked out of working hours.

## Removing temptation

1. Remove all ladders from scaffolds or securely board up the lower rungs to prevent access at the end of each working day.
2. Check that the perimeter hoarding or fencing is intact and secure.
3. Remove keys from plant and equipment when not in use.
4. Remove from view and secure any tools, equipment and materials that might tempt thieves onto the site after normal working hours.

## Dealing with trespassers

1. Ask suspected trespassers who they wish to see and, if necessary, escort them to site security. If they are genuine visitors, they will not mind being challenged.
2. Ensure that trespassing children are escorted to a place of safety immediately.
3. Do not put yourself in a position where you could be accused of assault.

Why should you prevent unauthorised access to the site?
How can you prevent the theft of tools, equipment and materials?
How can you remove the temptation for children to play on scaffold and in excavations?

Encourage a discussion by using a real-life situation or example and ask if there are any questions.

# Safety signs

| | |
|---|---|
| **Reason** | Safety signs are an important tool for making workers, visitors and others aware of hazards, information and emergency procedures on site. |
| **Why** | Employers must provide safety signs where there is a risk that cannot be avoided or controlled by another means. Safety signs are colour coded as required by The Health and Safety (Signs and Signals) Regulations and apply to all places of work. |
| **Outline** | This talk will cover safety signs you may come across whilst undertaking your job. |

### Example of prohibition sign

*No smoking*

### Example of mandatory sign

*Eye protection must be worn*

### Example of fire-fighting sign

*Emergency fire telephone*

### Example of warning sign

*Overhead load*

### Example of first aid sign

*Eyewash station*

## Types of safety signs

There are five types of safety signs:

- ✓ prohibition signs
- ✓ warning signs
- ✓ mandatory signs
- ✓ emergency escape or first aid signs
- ✓ fire-fighting signs.

# Safety signs

| | |
|---|---|
| **Reason** | Safety signs are an important tool for making workers, visitors and others aware of hazards, information and emergency procedures on site. |
| **Why** | Employers must provide safety signs where there is a risk that cannot be avoided or controlled by another means. Safety signs are colour coded as required by The Health and Safety (Signs and Signals) Regulations and apply to all places of work. |
| **Outline** | This talk will cover safety signs you may come across whilst undertaking your job. |

## Categories and what they mean

**Prohibition signs** are round and red, with a diagonal red line. These inform you of things you **must not do**. They stop the behavior likely to increase or cause danger, for example "No smoking" or "No access for pedestrians".

**Warning signs** are triangular and yellow with black edging. These warn you of a **hazard or danger**, for example 'toxic material' or 'overhead load'.

**Mandatory signs** are round and blue with a white pictogram. These inform you on things you **must do**, therefore prescribing a specific behaviour, for example 'safety gloves must be worn' or 'ear protection must be worn'.

**Emergency escape or first-aid signs** are rectangular or square and are green. These give you **information** on first aid, emergency exits or first-aid facilities, for example the location of the eye wash station or safety shower.

**Fire-fighting signs** are rectangular or square on a red background. They indicate the location of the facilities associated with fire safety, for example the location of the fire extinguisher or emergency fire telephone.

## Remember

1. Safety signs are only effective if you understand what they mean; if you do not understand a sign you should speak with your supervisor.
2. Safety signs must be visible. Do not place materials in the way or obstruct visibility. If you need to cover a sign, even temporarily, you should speak with your supervisor.
3. Check for safety signs when entering the site and at the start of every shift. New hazards may have been introduced to the area or the location of emergency equipment or the fire assembly point may have changed.
4. Ensure you wear all personal protective equiment (PPE) as identified on the mandatory signs on site.
5. Safety signs are there for you to obey. If you fail to obey a sign you could be putting yourself and others at risk and this could lead to disciplinary action.
6. If you introduce a new hazard to the area ensure you display the appropriate signage as required, so you can warn others of the hazard.

What colour are prohibition signs, and what do they mean?
If you see a sign which has been accidently covered by materials on site what should you do?
If you do not understand a sign what should you do?
Give an example of a mandatory sign.

Encourage a discussion by using a real-life situation or example and ask if there are any questions.
Ask for examples of other signs that are used on your site.

# Material storage

| | |
|---|---|
| **Reason** | Unsafe stacking can lead to serious injuries. |
| **Why** | You don't want to find yourself underneath an unsafe stack. |
| **Outline** | This talk covers general points on stacking bricks, timber, pipes and prefabricated panels. |

*Always stack safely*

| Notes |
|---|
| |
| |
| |
| |
| |
| |
| |
| |

Stack safe – stay safe.

# Material storage

| | |
|---|---|
| **Reason** | Unsafe stacking can lead to serious injuries. |
| **Why** | You don't want to find yourself underneath an unsafe stack. |
| **Outline** | This talk covers general points on stacking bricks, timber, pipes and prefabricated panels. |

## Stacking

1. When handling materials, wear work gloves and safety boots.
2. Only stack material in authorised areas; never near doorways, accessways or on fire escape routes.
3. Stack on a level surface and provide packing.
4. Never make stacks higher than three times the minimum base width.
5. Stack close to the work area to reduce the amount of handling.
6. If material is being lowered by machine, keep hands clear of the load.

## Bricks, blocks and palleted material

1. Ensure the base of the stack is level. Only stack two packs high.
2. Ensure the upper pack is loaded squarely on to the lower one.
3. If banding is damaged or materials are displaced in the pack, do not stack other materials on top.

## Timber

1. Racks are the best method of storing small-sized timbers.
2. Joists and larger timbers should be placed on bearers.
3. Try to keep different lengths in separate stacks.

## Large prefabricated panels

1. Stack flat or store secured in designed racks (stillages).
2. Don't lean against parts of semi-constructed buildings or temporary structures.
3. Don't store upright where panels can be affected by wind.
4. Secure if stored at height or in other exposed areas.

## Pipes and tubes

1. Where pipes are small in diameter stack in racks, if large in diameter securely chock at the base.
2. Don't stack in pyramids as they can become unstable.
3. Large concrete rings must be laid flat so that they cannot be moved or rolled by any person, especially children.

?

Where should materials never be stacked?
What should you consider before loading material in a stack?
If you see damaged banding, what should you NOT do?
What should be used when stacking larger timbers?
Where and how should panels be stacked?
How should you secure large diameter pipes?
How can you reduce the amount of handling?
Where should you not stack large panels?

Encourage a discussion by using a real-life situation or example and ask if there are any questions.

# Slips, trips and falls (Housekeeping)

| | |
|---|---|
| **Reason** | Every year, many injuries occur through slips, trips and falls. |
| **Why** | Most of these injuries are easily preventable with a little care. |
| **Outline** | This talk covers the causes and prevention of slips, trips and falls. |

*In these photos, leads are kept above head height to avoid tripping hazards.*

Tidy up as you go. Your carelessness could cause serious injuries to someone else.

# Slips, trips and falls (Housekeeping)

| | |
|---|---|
| **Reason** | Every year, many injuries occur through slips, trips and falls. |
| **Why** | Most of these injuries are easily preventable with a little care. |
| **Outline** | This talk covers the causes and prevention of slips, trips and falls. |

## Why do they occur?

1. Most injuries from slips, trips and falls occur because of poor housekeeping.
2. Many items left on the ground (such as coiled cables, hand tools and lengths of pipe or timber), will trip someone if not moved to a safe position.
3. Spilt substances (such as oils and greases) will form a slip hazard if not immediately cleaned up.
4. General debris (such as brick and block fragments) can quickly accumulate and form a tripping hazard.
5. Trailing cables are another frequent cause of tripping.
6. Mud left on the rungs of a ladder will be a slipping and falling hazard for the next person.
7. Reduced levels of natural light (for example, during winter afternoons) can easily increase the tripping hazards if adequate access lighting is not provided. Tools, equipment and materials that are visible in full daylight might be hidden in semi-darkness.

## What you must do

1. Clear up waste materials as you create them. Lightweight waste should be bagged or bundled, and nails removed from waste timber.
2. Work in a tidy manner: do not leave tools, equipment or unused materials lying about on the floor.
3. If you are using substances that could spill, ensure that you have a means of effectively clearing up the spillage.
4. As far as possible, route cables for power tools above head height. If cables have to be routed at floor level, try to avoid crossing pedestrian walkways.
5. If the site is muddy, scrape mud off your boots before climbing ladders or walking anywhere else where it might be a danger to others.
6. Be aware of the increased risks of tripping as the level of natural light fades; ensure that all tools, equipment and materials are stored in a safe location.
7. Tell your supervisor if you do not have the means available to clear up and dispose of the waste you create.

?

What can you do in your job to reduce tripping hazards?
What would be an effective way of clearing up spilt liquid?

Encourage a discussion by using a real-life situation or example and ask if there are any questions.

# Fire

| | |
|---|---|
| **Reason** | Fire kills on average 400 people in Britain every year and injures thousands more. |
| **Why** | Knowing how to prevent a fire may save lives. |
| **Outline** | This talk covers fire prevention, precautions, types of extinguisher and actions on fire. |

### Types of portable fire extinguisher

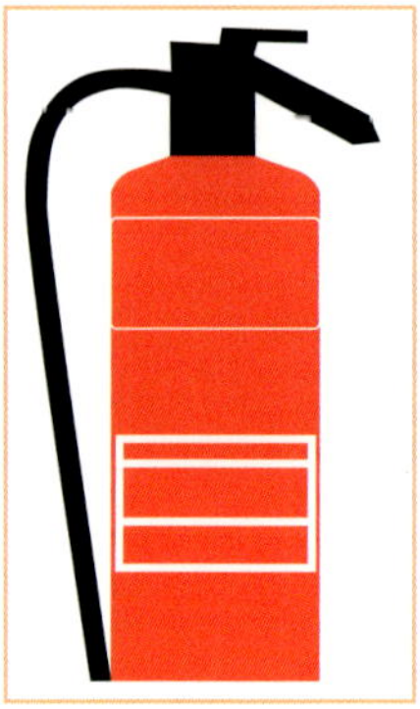

*Water (red label)*

*Foam (cream label)*

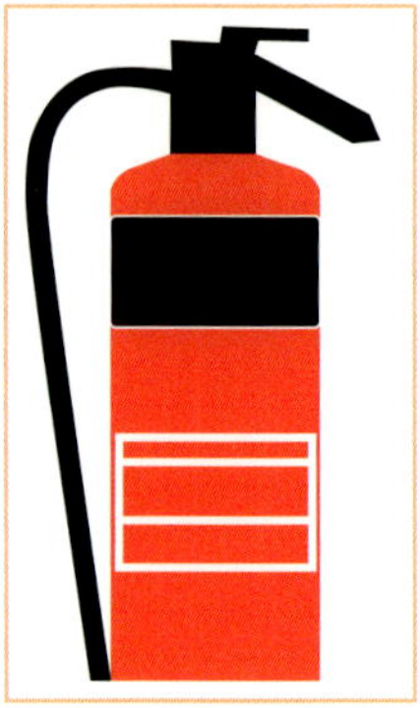

*Carbon dioxide (black label)*

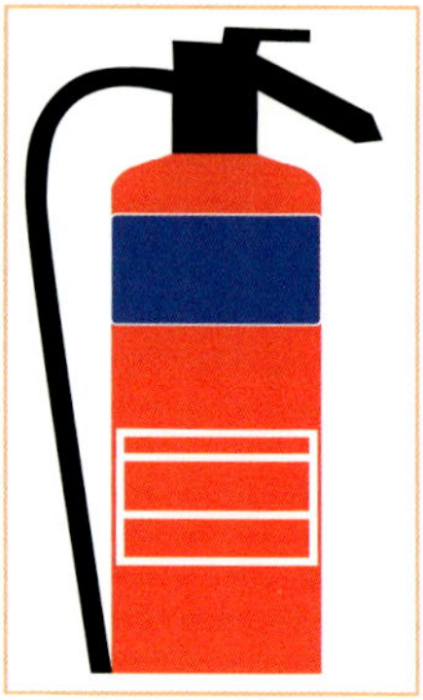

*Dry powder (blue label)*

Plan in advance. You won't have time when fire breaks out.

# Fire

| | |
|---|---|
| **Reason** | Fire kills on average 400 people in Britain every year and injures thousands more. |
| **Why** | Knowing how to prevent a fire may save lives. |
| **Outline** | This talk covers fire prevention, precautions, types of extinguisher and actions on fire. |

## Fire prevention

1. Don't hang clothing over or near heating equipment.
2. Don't let paper, oily rags or other rubbish accumulate and don't smoke in prohibited areas.
3. Use purpose-made containers when handling or storing flammable liquids, not open tins or buckets.
4. Don't overload electric sockets – one socket, one plug.
5. Handle flammable liquids at a safe distance from possible sources of ignition.
6. Ensure there are no adjacent flammable materials before using blowlamps and cutting equipment.
7. Bitumen boilers, soldering irons and gas rings must be on non-combustible stands.
8. When electrical equipment is not in use, switch it off: beware of heat produced by halogen lamps.

## Fire precautions

1. Make sure you know what to do in case of fire.
2. Make certain you know your escape route and assembly point.
3. Keep escape routes clear and unobstructed.
4. Don't obstruct access to fire-fighting equipment.
5. Only attempt to fight a fire if you have been trained to select and use a fire extinguisher.

## Fire extinguishers

1. All extinguishers should now be coloured red with a contrasting colour panel to indicate the contents.
   - **Water** (red) for use on paper, wood and solid flammables.
   - **$CO_2$** (black) for use on liquids, gases and electrical fires.
   - **Foam** (cream) for use on flammable liquids and solid flammables.
   - **Powder** (blue) can be used on all types of fires but is not ideal outdoors in windy weather.

## In the event of a fire

1. Raise the alarm and then call the fire brigade.
2. Close doors and windows to prevent the spread of fire.
3. Evacuate the building or area you are working in.
4. If trained, fight the fire with extinguishers provided. Don't put yourself at risk. Always ensure you have an escape route.

What should flammable liquids be stored in?
What checks should you carry out before and after using cutting and welding equipment?
Name the precautions concerning extinguishers and fire exits.
What type of fire can water be used on?
What actions should you take in the event of a fire?
What extinguisher must be used on flammable liquids?
List five ways of preventing fires on site.

Encourage a discussion by using a real-life situation or example and ask if there are any questions.

# LPG and other compressed gases

| | |
|---|---|
| **Reason** | Liquefied petroleum gas (LPG) and other compressed gases, if used safely, are a convenient and valuable source of energy. |
| **Why** | If you know how to use it safely you can prevent serious accidents and injury. |
| **Outline** | This talk covers hazards, use, storage and transportation. |

*Store LPG appropriately*

*Smoking is prohibited near LPG storage*

Compressed gas and LPG are useful tools, but potentially lethal ones.

# LPG and other compressed gases

| | |
|---|---|
| **Reason** | Liquefied petroleum gas (LPG) and other compressed gases, if used safely, are a convenient and valuable source of energy. |
| **Why** | If you know how to use it safely you can prevent serious accidents and injury. |
| **Outline** | This talk covers hazards, use, storage and transportation. |

## Hazards

1. Treat every cylinder as being full and handle carefully.
2. Keep cylinders away from sun, artificial heat, flammable materials, corrosive chemicals and fumes.
3. Avoid damage to valves and fittings. Don't use them for lifting or carrying.
4. When using mixed gases (such as welding), flashback arrestors must be fitted.
5. Don't smoke when using compressed gases.
6. Don't use cylinders as rollers for moving equipment.
7. If gas cylinders are being heated in a fire, call the fire brigade and evacuate the area.

## Safe use

1. Regular inspection of hoses, cylinders and valves should be undertaken before use.
2. Open cylinder valves slowly and close sufficiently to shut off gas – never use force.
3. Ensure that gloves (if worn) are free from oil and grease.
4. Keep valves and fittings of oxygen cylinders free from oil and similar items.
5. Keep gas hoses clear of traffic to prevent them being damaged.
6. Know what to do in an emergency.
7. Ensure a fire extinguisher is always available for hot work but only try to fight secondary fires.
8. If there's a hot-work permit or procedure, follow it.

## Storage and transportation

1. Always secure acetylene cylinders in an upright position.
2. Store all cylinders so that they cannot fall or roll.
3. Always lift cylinders from trucks, don't drop or slide them.
4. Move full-size cylinders using a trolley. If one is not available, get assistance.
5. Transport cylinders in vehicles with good ventilation.

What should not be used for carrying a cylinder?
If a cylinder is being heated by a fire, what is the hazard?
What should be done before using compressed gases?
What must you remember when working close to traffic?
How should cylinders be stored and transported?
What cylinder must be kept free of oil and grease?
How must cylinder valves be opened?

Now inform your workforce of the company policy on hot-work permits or procedures.
Encourage a discussion by using a real-life situation or example and ask if there are any questions.

# Petrol and diesel

| | |
|---|---|
| **Reason** | The amount of plant operating on most sites will result in the storage and use of large quantities of fuel, which creates hazards. |
| **Why** | By being aware of the risks, the chance of an accident occurring can be reduced or eliminated. |
| **Outline** | This talk covers the hazards created by different fuels and the precautions necessary. |

*5-litre plastic diesel can*

*5-litre plastic petrol can*

*20-litre petrol jerry can*

*205-litre diesel drum*

Do not bring petrol into timber frames or enclosed buildings – use designated refueling points or better still another type of powered equipment.

# Petrol and diesel

| | |
|---|---|
| **Reason** | The amount of plant operating on most sites will result in the storage and use of large quantities of fuel, which creates hazards. |
| **Why** | By being aware of the risks, the chance of an accident occurring can be reduced or eliminated. |
| **Outline** | This talk covers the hazards created by different fuels and the precautions necessary. |

## Petrol

1. Usually, only small plant (such as disc-cutters and chainsaws) now run on petrol.
2. Petrol fumes are highly flammable – only refuel plant in well ventilated areas.
3. Do not store excessive quantities of petrol.
4. Petrol must only be stored in purpose-designed containers – 10 litres maximum (five litres in a plastic container).
5. No smoking or other sources of ignition are allowed in areas where petrol is stored or decanted.

## Diesel

1. There are no storage restrictions for diesel fuel.
2. Protective gloves should be worn when handling diesel oil because skin contact can result in irritation.
3. As an oil, spilt diesel will cause a slip hazard on hard surfaces.
4. Diesel oil should be stored in metal cans, which should be kept in a lockable store.

## Liquefied petroleum gas (LPG)

1. LPG is used mainly as a fuel for small plant vehicles (such as dumpers and forklift trucks).
2. Cylinders are of special construction and designed to be mounted on their side.
3. Cylinder connectors and other unions have a left-hand thread.
4. Use the correct size spanner for tightening connections; hand-tight connections will leak.
5. LPG vapour is heavier than air; leaks will accumulate at floor level or in drains if not allowed to disperse.
6. LPG vapour is highly flammable and must be kept away from sources of heat, naked flames and sparks.
7. If LPG cylinders are being heated in a fire, evacuate the area.

## Fuel storage

1. Storage areas must be secure, well ventilated and away from sources of ignition.

How do you reduce the chance of diesel oil spillage when refuelling plant?
What arrangements should be made for the storage of LPG cylinders?
How can you prevent the accumulation of leaking LPG at floor level?
What type of fire extinguisher should you have for hot work using LPG?

Now inform your workforce of the company policy regarding the safe storage of vehicle fuels.
Encourage a discussion by using a real-life situation or example and ask if there are any questions.

# Underground and overhead services

| | |
|---|---|
| **Reason** | Unseen, unheard, electricity can cause death or serious injury without warning. |
| **Why** | Your body is an extremely good conductor of electricity – don't find out the hard way. |
| **Outline** | This talk covers underground cables and overhead power lines. |

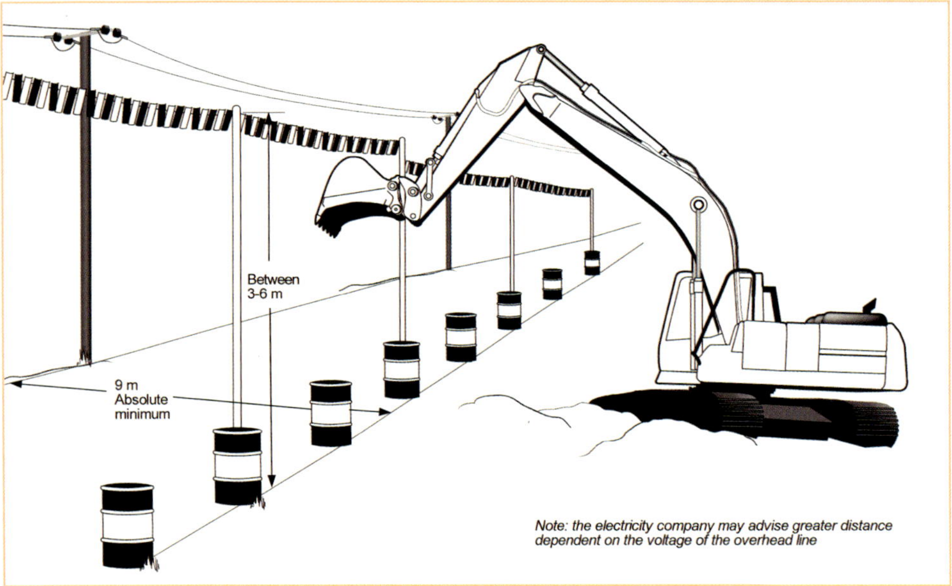


*If mobile cranes or excavators are used, the minimum distance from the ground level barrier to the line should be 9 m if on wood or metal poles, 15 m if on pylons, PLUS the length of the jib or boom*

Don't let a 'live' take a 'life'. Look out, look up.

# Underground and overhead services

| | |
|---|---|
| **Reason** | Unseen, unheard, electricity can cause death or serious injury without warning. |
| **Why** | Your body is an extremely good conductor of electricity – don't find out the hard way. |
| **Outline** | This talk covers underground cables and overhead power lines. |

## Underground cables

1. Before digging, check plans provided by the electricity company, telephone and cable TV company.
2. Before digging, use a cable-locating device that is in good working order. Ensure you are trained to use it.
3. Assume all cables are live, unless your supervisor tells you they are dead.
4. Hand dig trial holes to expose cables, look for marker tape or tiles above the cables. Continue using the cable locator to establish exact location.
5. Don't assume that a buried cable will run in a straight line between two known points.
6. When exposed, protect the cable from damage and support it.
7. If a cable is accidentally damaged, keep everyone clear until the owner has been told and inspected it.
8. During back-filling, ensure marker tapes or tiles are replaced.
9. If using power tools to break up concrete surfaces, avoid overpenetration as cables may be directly underneath.

## Overhead power lines

1. Until it has been proved otherwise, all overhead lines should be treated as live and dangerous.
2. Ensure you know the maximum clearance distances specified by the electricity supply company.
3. Do not bypass goal posts, barriers or other warnings.
4. Check your route is clear of overhead power lines before moving a mobile scaffold tower or metal ladder.
5. If signalling, always keep power lines in view. Guide plant under power lines where goal posts have been erected.
6. Ensure you observe special precautions laid down by the electricity company before working under overhead lines.
7. If erecting scaffolding adjacent to power lines, ensure the poles are handled a safe distance away.
8. Don't stack materials or operate tippers under power lines as it will reduce the safe clearance and can result in arcing.

?

Before digging, name two things that must be done.
What action should you take if you accidentally damage an underground electric cable?
Who should you ask what the safe working distance is?
When erecting scaffolding close to power lines, what precautions must you take?
What must you remember during back-filling?
As a signaller, what must you always keep in view?

Encourage a discussion by using a real-life situation or example and ask if there are any questions.

# Plant and equipment

| | |
|---|---|
| **Reason** | The misuse of plant or lack of maintenance can lead to injuries to the user and others. |
| **Why** | Operators of power-operated plant and equipment must be trained in its use and authorised. |
| **Outline** | This talk covers safe operating procedures for plant and equipment. |

*Secure charging station*

*Protecting cables*

| Notes |
|---|
| |
| |
| |
| |
| |
| |
| |
| |

# Plant and equipment

| | |
|---|---|
| **Reason** | The misuse of plant or lack of maintenance can lead to injuries to the user and others. |
| **Why** | Operators of power-operated plant and equipment must be trained in its use and authorised. |
| **Outline** | This talk covers safe operating procedures for plant and equipment. |

## Definition

1. Plant and equipment can be mobile or static equipment used in construction.
2. Examples include dumper trucks, cement mixers, bar-bending machines and welding sets.

## General precautions

1. Plant and equipment should only be used by people who have been trained in their specific use and are authorised.
2. Consider the risks to other people who are nearby when operating plant and equipment.
3. Before use, ensure that plant and equipment have no obvious defects; bring defects to your supervisor's attention.
4. Only use plant and equipment for its intended purpose.
5. Be aware of any pre-use checks and carry them out.

## Mobile plant and equipment

1. Don't carry passengers unless the plant is designed to do so and wear seat-belts where fitted.
2. Observe site speed limits and one-way systems and, if necessary, obtain assistance when reversing.
3. Carry out daily checks (brakes, oil, lights and tyres).
4. Be cautious of crush-zones on rear-wheel steer and centre-pivoting plant.
5. If necessary, vacate the driver's seat during loading/unloading.

## Static plant and equipment

1. If fitted with wheels, ensure brakes are on or wheels are securely chocked.
2. If engine-driven, ensure exhaust gases cannot accumulate.
3. If electrically-powered, ensure supply cable and plug cannot be damaged.
4. Ensure all guards are in position.
5. Consider the need for barriers around the equipment to protect others.

What should happen before anyone is allowed to operate plant or equipment?
What are the usual pre-use checks?

Encourage a discussion by using a real-life situation or example and ask if there are any questions.

# Abrasive wheels

| | |
|---|---|
| **Reason** | Many accidents involve the wrong type of abrasive wheel being fitted or machines being used by untrained operators. |
| **Why** | Without the correct knowledge more accidents will continue to occur. |
| **Outline** | This talk covers general precautions when using portable and bench-mounted abrasive wheels, and floor-mounted woodworking machines. |

*Abrasive wheel, too small due to excessive use, and a snagging hazard*

*Failure due to poor storage or age of disc*

*Extreme thermal damage caused by faulty machine bearings*

*Damage caused by poor machine maintenance*

*Damage caused by heavy use*

*Damage caused by poor manufacturing quality*

Abrasive wheels must only be fitted and used by a trained and competent person.

# Abrasive wheels

| | |
|---|---|
| **Reason** | Many accidents involve the wrong type of abrasive wheel being fitted or machines being used by untrained operators. |
| **Why** | Without the correct knowledge more accidents will continue to occur. |
| **Outline** | This talk covers general precautions when using portable and bench-mounted abrasive wheels, and floor-mounted woodworking machines. |

## General precautions

1. The speed of the machine must not exceed the maximum permissible speed of the wheel. Many accidents are caused by the wheel overspeeding.
2. Don't mount an abrasive wheel unless you are authorised in writing and trained to do so.
3. Don't exert heavy pressure on the wheel.
4. Never use the side of the wheel unless it is designed for it.
5. Hearing and eye protection must always be worn.
6. Where practical use wet-cutting to reduce dust.

## Safe use of portable abrasive wheels

1. Only use reinforced discs on hand-held machines.
2. Adjust the guard to expose the minimum wheel surface necessary for the operation.
3. Be aware of other workers in your area and don't put them at risk by your actions.
4. Depending upon what is being cut, it is likely that appropriate respiratory protective equipment (RPE) will be needed.
5. Be aware of noise and vibration hazards.

## Safe use of bench-mounted abrasive wheels

1. Adjust the tool rest as close as possible to the face of the wheel and keep the glass screen in the safety position.
2. Keep your fingers below the tool rest level.
3. Use the correct grade of wheel for the work in hand.
4. Keep the face of the wheel evenly dressed and don't use the side of the wheel.
5. After fitting, run a replacement wheel for a full minute before attempting to use it. Stand clear during the test.
6. Stop the wheel when not in use.
7. It is likely that RPE or extraction will be necessary.

Why is the rated speed of the wheel important?
What type of discs must be fitted to portable machines?
Where should the tool rest be positioned?
For how long should a replacement wheel be run before using it?
What must be worn when using abrasive wheels?
Which part of a wheel should not be used?
Who should mount and adjust abrasive wheels?

Now inform your workforce of the company policy regarding working with abrasive wheels.
Encourage a discussion by using a real-life situation or example and ask if there are any questions.

# Diamond blades

| | |
|---|---|
| **Reason** | Diamond blade cutting is potentially a dangerous activity if not competently controlled. |
| **Why** | Knowing and understanding the risks, and taking the proper precautions, will keep you and the workplace safe. |
| **Outline** | This talk covers the main points on selection, hazards and common injuries when using diamond blades. |

*Wear the appropriate PPE and ensure the guards are properly set*

*Wet dust suppression reduces exposure to silica dust*

# Diamond blades

| | |
|---|---|
| **Reason** | Diamond blade cutting is potentially a dangerous activity if not competently controlled. |
| **Why** | Knowing and understanding the risks, and taking the proper precautions, will keep you and the workplace safe. |
| **Outline** | This talk covers the main points on selection, hazards and common injuries when using diamond blades. |

## General precautions

1. Complete a risk assessment to minimise hazards (such as dust (especially silica dust), noise and vibration) caused to the operator, workers and general public who are nearby, or consider other methods of completing the work.
2. Manufacturer's instructions must be followed when selecting the correct blade for the task.
3. Diamond blades are very strong and are long-lasting if used correctly.
4. Blades should only be mounted by a trained, competent and authorised person.
5. Cutting tools should only be used by trained and competent people.
6. Wet-cutting removes the problem of dust and reduces wear by preventing it from getting into the machine. However, it creates slurry that needs managing, as well as wetting the operator.

## Hazards

1. The operator's hand must not come into contact with the revolving wheel.
2. Beware of particles (such as hot metal or sparks) being thrown off while in use.
3. Dust from dry-cutting (especially if there is silica content) can cause respiratory problems.
4. Beware of disintegration (or loss of segments) of an overspeeding, damaged or incorrectly used blade.

## Common causes of injuries

1. Using the wrong type of blade or incorrect mounting.
2. Fire hazards from sparks flying off the blade or the material being cut overheating.
3. Diamond blades losing segments.
4. Damage to parts of the body by not using the correct PPE, poor posture, an obstructed workplace and unsecured or out of balance/unstable material to be cut.

What safety precautions should you consider when cutting kerbstones, paving slabs and concrete blocks?
Given the choice of wet or dry cutting which would you prefer and why?
What PPE should you wear?
Why is wind direction important if you have to dry cut?

Now inform your workforce of the company policy on mounting and using disc-cutters and saws.
Encourage a discussion by using a real-life situation or example and ask if there are any questions.

# Cartridge-operated tools

| | |
|---|---|
| **Reason** | Cartridge-operated tools must be used in a safe manner. |
| **Why** | If cartridge-operated tools are used recklessly or incompetently they can be lethal. |
| **Outline** | This talk covers what you should do before use, the hazards and how to use these tools safely. |

*Cartridge tool*

*Cartridge shots*

*Cartridge tool shots*

*Using a cartridge tool on a stable surface*

*Safe storage of a cartridge tool*

Always treat cartridge-operated tools with respect.

# Cartridge-operated tools

| | |
|---|---|
| **Reason** | Cartridge-operated tools must be used in a safe manner. |
| **Why** | If cartridge-operated tools are used recklessly or incompetently they can be lethal. |
| **Outline** | This talk covers what you should do before use, the hazards and how to use these tools safely. |

## Before use

1. You must be trained, over 18, with a certificate of authority.
2. Read the maker's instructions carefully before using the tool.
3. Load the tool with the barrel pointing away from you.
4. Never walk around with a loaded tool. Load it just before use.

## Hazards

1. Using a cartridge that is too powerful for the task.
2. Voids in the structure being fired into or the material being fired into is too thin.
3. Firing into the hole of a previously attempted fixing.
4. Trying to fix into excessively hard material.
5. The tool is not held square to the surface, or the fixing is too near the edge of the material.

## Safe use

1. Always wear PPE when using cartridge-operated tools.
2. Hold the tool at right angles to the job when firing.
3. Check the material into which the fixing is to be fired. Carry out a test fire first. Check that no-one is behind the target.
4. Allow at least 75 mm from edges of concrete or brickwork.
5. Ensure the complete splinter guard is resting on the work surface.
6. Never place your hand over the end of the barrel.
7. In the event of a misfire, wait a minute without withdrawing the tool, then refire it. If nothing happens, wait a further minute before unloading.

## After use

1. Keep the tool clean and well oiled.
2. Never leave the tool loaded when not in use.
3. Cartridges must be kept under lock and key in a safe place.

Before using a cartridge tool, what should you do?
What is through penetration caused by?
What are the main reasons for ricochets?
How far from the edge of concrete should the tool be?
What should you do with the tool after use?
If you have a misfire, what should you do?
Before firing into a target, what should you check for?

Now inform your workforce of the company policy regarding cartridge-operated tools.
Encourage a discussion by using a real-life situation or example and ask if there are any questions.

# Compressed gas tools

| | |
|---|---|
| **Reason** | Compressed gas tools can be lethal in untrained hands. |
| **Why** | Awareness of the hazards and the consequence of improper use is essential for the safety of the user and others within the work area. |
| **Outline** | This talk covers pre-use checks, safe use, PPE and work area protection. |

*Incorrect storage of a gas tool*

*Cartridges*

| Notes |
|---|
| |
| |
| |
| |
| |
| |
| |
| |
| |
| |

# Compressed gas tools

| | |
|---|---|
| **Reason** | Compressed gas tools can be lethal in untrained hands. |
| **Why** | Awareness of the hazards and the consequence of improper use is essential for the safety of the user and others within the work area. |
| **Outline** | This talk covers pre-use checks, safe use, PPE and work area protection. |

## Compressed gas tools

1. Compressed gas tools are the modern equivalent of cartridge-operated tools and have similar hazards associated with them.
2. The modern-day compressed gas tool could be described as a machine gun, compared to a single-shot weapon like a blunderbuss.
3. Gas cartridges are used as the propellant and are often ignited via a trigger action with the use of batteries.
4. There are many variations that have magazines for nails and fixings, in a range of shapes and sizes, with penetration depths suitable for items such as softwood frames or concrete.

## Before use

1. Do not remove the tool from its box unless you have been trained to use it.
2. Follow the manufacturer's instructions for the specific tool type and, if in any doubt, ask for help.
3. Be aware of the type of surface that you are fixing to and check for hidden voids.
4. Ensure that there are no buried or hidden services that you could hit.
5. On refurbishment projects information about existing services may not be available or, if it is, it may not be accurate.

## Protective measures

1. Ensure that any compressed gas tool is properly maintained and protected from damage.
2. Use the correct PPE, as specified by both the manufacturer and your employer.
3. If you discover a malfunction or any faults with the equipment stop work immediately and report it to your supervisor.

What PPE should you wear when you use a compressed gas tool?
What pre-user checks should you carry out before using a compressed gas tool?
Why is it important to check the cartridge on the tool with the surface to be nailed?

Encourage a discussion by using a real-life situation or example and ask if there are any questions.

# Woodworking machines

| | |
|---|---|
| **Reason** | Woodworking machines can be particularly dangerous if they are not used properly. |
| **Why** | In order to function, at least a part of the blade must be exposed during use. |
| **Outline** | This talk covers safety precautions in the use of hand-held and floor-mounted woodworking machines. |

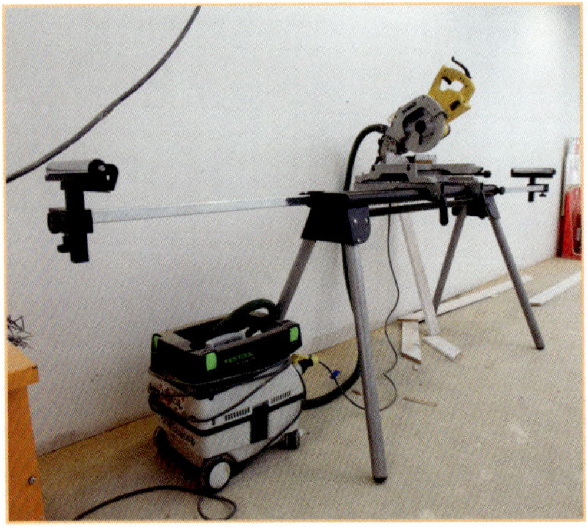

*You may need training before using powered woodworking tools*

# Woodworking machines

| | |
|---|---|
| **Reason** | Woodworking machines can be particularly dangerous if they are not used properly. |
| **Why** | In order to function, at least a part of the blade must be exposed during use. |
| **Outline** | This talk covers safety precautions in the use of hand-held and floor-mounted woodworking machines. |

## Types of woodworking machine

1. Woodworking machines include fixed and portable machines.
2. Examples include circular saws, bandsaws, routers, planers and thicknessers.

## General precautions

1. Most woodworking machines are electrically powered, so ensure that the electrical supply cable is in good condition and out of harm's way.
2. Consider the risks to other people who are nearby when operating woodworking machines.
3. Do not operate any woodworking machine unless you have been trained to do so.
4. Ensure all fixed guards are in position.
5. Ensure all pivoting guards are free to move and properly positioned during use.
6. Ensure that you have the correct PPE before starting work.
7. Ensure that you are not wearing any loose clothing.
8. Promptly clean up timber offcuts from the floor. Do not leave them as a tripping hazard.
9. Ensure that the supply is securely locked off and that you retain the key during maintenance.

## Hand-held circular saws

1. Reduce the chance of fatal electric shock by using 110 volt tools.
2. Ensure the power lead is disconnected from the supply before adjusting the saw.
3. Adjust the depth and angle of cut, and the position of the fence before use.
4. Ensure dust extraction adaptors and bags are fitted and used, if available.
5. Ensure the blade is securely fixed, sharp and not cracked.

## Floor-mounted woodworking machines

1. Check that the start and stop controls can be operated without danger.
2. Ensure that the cutter or blade is guarded to the greatest practical extent.
3. Check that there is sufficient working space around each machine.
4. Ensure that the level and direction of lighting is satisfactory for safe working.
5. Use the dust and chip extraction system, where fitted.
6. Check that machines are level and securely fixed down for stability.

Why should dust extraction equipment be fitted?
What can you do to ensure the safety of others whilst using the saw?
What should be used to feed timber if there is no motorised feed?

Encourage a discussion by using a real-life situation or example and ask if there are any questions.

# Portable, hand-held electrical tools

| | |
|---|---|
| **Reason** | Electrical tools face harsh conditions on site and when misused they get damaged and become dangerous. |
| **Why** | In one year, there were 194 reported incidents of electric shock involving portable electrical tools. |
| **Outline** | This talk covers pre-use checks, use of electrical tools and hazards. |

*Wear correct PPE*

*Portable powered tools should be in good condition*

Look after portable electrical tools and they will look after you.

# Portable, hand-held electrical tools

| | |
|---|---|
| **Reason** | Electrical tools face harsh conditions on site and when misused they get damaged and become dangerous. |
| **Why** | In one year, there were 194 reported incidents of electric shock involving portable electrical tools. |
| **Outline** | This talk covers pre-use checks, use of electrical tools and hazards. |

## Pre-use checks

1. Make sure the casing isn't damaged and, if it is, don't use it.
2. Make sure that all cables, plugs or connectors are sound and not damaged.
3. Use tools on the correct power supply. Generally only 110 volt tools are permitted on site.
4. Check for a current portable appliance testing (PAT) test sticker.

## Safe use

1. Portable electrical tools should only be used for their designated purpose.
2. Ensure switches are working correctly before connecting to the power supply.
3. Wear eye protection if there is any risk to your eyes.
4. Disconnect tools when not in use or when making adjustments.
5. Electrical power tools should be regularly inspected and maintained by a competent electrician.

## Hazards

1. If possible, keep power cables off the floor. They may get damaged and cause someone to trip, or they may trail through water.
2. Electrical tools often present a noise hazard. Wear hearing protection if necessary.
3. Keep equipment clean and dry.
4. Portable electrical tools that have become wet should be allowed to dry and then be checked by a competent person for electrical safety before being used again.
5. Some hand-held tools can cause hand-arm vibration.

What two things should you check before using a portable electrical tool?
When drilling, when should you wear eye protection?
What are three potential hazards when using portable electrical tools?
What voltage tools should be used on site?
Who should inspect and maintain portable electrical tools?

Encourage a discussion by using a real-life situation or example and ask if there are any questions.

# Site transport and pedestrian segregation

| | |
|---|---|
| **Reason** | Site transport carries people and materials. The carriage of both should be carried out in safety. |
| **Why** | Many accidents have occurred in the past because site transport was operated in an unsafe manner. |
| **Outline** | This talk covers the dangers associated with site transport and how they can be overcome. |

*Examples of good pedestrian segregation*

# Site transport and pedestrian segregation

| | |
|---|---|
| **Reason** | Site transport carries people and materials. The carriage of both should be carried out in safety. |
| **Why** | Many accidents have occurred in the past because site transport was operated in an unsafe manner. |
| **Outline** | This talk covers the dangers associated with site transport and how they can be overcome. |

## General precautions

1. Never be tempted to drive site transport unless you have been properly trained. It is recommended that you hold an appropriate competency card, for example a CSCS card, if necessary an appropriate class of driving licence and have been authorised to drive the site transport.
2. Drivers of site transport should carry out daily pre-use checks of their vehicles and report any defects found.
3. People are injured or killed by site transport that is reversing without the assistance of a signaller.
4. The very nature of some site transport means that the driver has only limited visibility from the driving position.
5. Drivers should obey site speed limits and one-way systems.
6. When parking, ensure the parking brake is on and the wheels are chocked if necessary. Accidents have been caused by items of runaway site transport.
7. If site transport is left after working hours, ensure it is immobilised and in a safe state especially where children might congregate.
8. Use stop-blocks where provided to prevent over-running.
9. Site transport should only be refuelled at designated refuelling points by trained operatives.

## Site transport for carrying materials

1. Site transport used for carrying materials must not be overloaded; drivers must know the maximum safe load.
2. Loads that could fall off must be adequately secured.
3. Ideally, sites will be organised so that the need to reverse is kept to a minimum; if so, do not reverse unnecessarily.
4. If you are involved in, or working near, tipping operations, keep well clear whilst materials are actually being tipped.
5. Stay well clear of the unpropped bodies of tipper lorries.
6. Site transport intended for carrying materials must not carry passengers unless it is designed to do so.

## Site transport for carrying people

1. Drivers of people-carrying site transport have a particularly important responsibility; theirs is a valuable cargo.
2. In many cases, it will be advisable to exclude people-carrying site transport, such as crew buses, from all but access roads.
3. People-carrying site transport is more likely to travel on public highways and so must comply with relevant legislation.
4. Passengers on people-carrying site transport must always act in a responsible manner and not endanger the vehicle.

How is limited visibility on some vehicles overcome?
What should you do if you are asked to drive an item of site transport that you have not been trained on?
What precautions should be taken to prevent environmental pollution at refuelling points?

Now inform your workforce of company policy for ensuring the competency of site transport drivers.
Encourage a discussion by using a real-life situation or example and ask if there are any questions.

# Mobile plant

| | |
|---|---|
| **Reason** | The use of mobile plant can be lethal in untrained hands. |
| **Why** | Safe operation of mobile plant requires competence; a combination of training and experience. |
| **Outline** | This talk covers the dangers associated with mobile plant and how they can be overcome. |

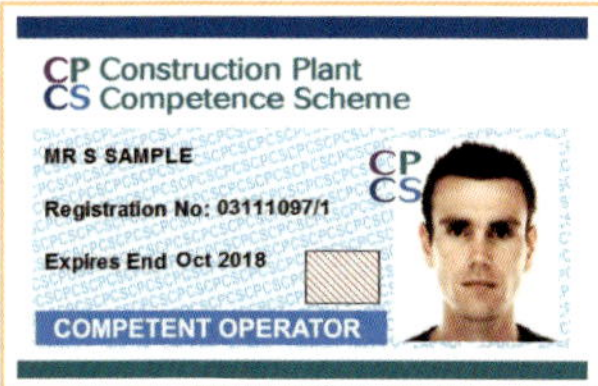


*Example of a CPCS blue card*

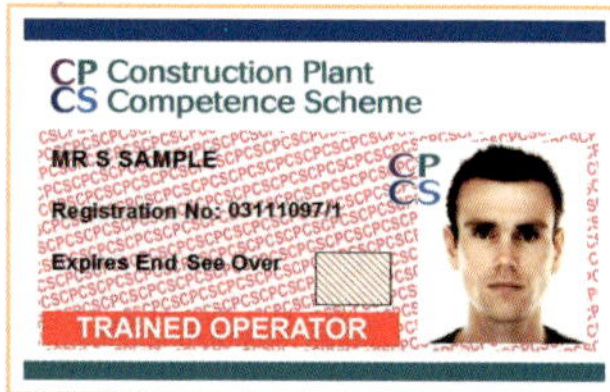


*Example of a CPCS red card*

*Only operate plant if you are trained and competent*

*Taking risks or being complacent can have serious consequences*

Mobile plant is there to help you. Don't let it harm you.

# Mobile plant

| | |
|---|---|
| **Reason** | The use of mobile plant can be lethal in untrained hands. |
| **Why** | Safe operation of mobile plant requires competence; a combination of training and experience. |
| **Outline** | This talk covers the dangers associated with mobile plant and how they can be overcome. |

## The dangers

1. Certain items of plant (such as dumpers) are particularly prone to overturning if used unsafely.
2. Groundwork and lifting activities account for the greatest number of accidents involving mobile plant.
3. Accidents are caused by plant being used unsafely by untrained, unauthorised workers.
4. People are injured or killed by mobile plant that is reversing without the assistance of a signaller.
5. On some mobile plant the driver has only limited visibility from the driving position.

## General precautions for drivers

1. Never use mobile plant unless you have been trained, hold the necessary competency card, such as a CPCS card and have been authorised.
2. Operators of mobile plant should carry out daily pre-use checks and report any defects noticed.
3. In most cases it will be necessary for a signaller to assist the driver during reversing.
4. Ideally, sites will be planned so that the need to reverse is kept to a minimum; if so, do not reverse unnecessarily.
5. Obey site speed limits and one-way systems.
6. When parked, ensure the parking brake is on and the wheels are chocked if necessary.
7. If mobile plant is left where children might congregate, ensure it is immobilised and in a safe state.
8. Use stop-blocks where provided to prevent over-running.
9. Mobile plant should only be refuelled at designated refuelling points.
10. Don't carry passengers unless the plant is designed to do so.
11. If a roll over protection system (ROPS) is not provided, vacate the driver's seat during loading/unloading.

## General precautions for others

1. On some sites, the routes of mobile plant and private vehicles may cross; make sure you know who has priority.
2. Keep well away from operating mobile plant; on well organised sites, mobile plant and pedestrians are segregated.
3. If your safety is at risk from any item of mobile plant, stop work and report it to your supervisor.
4. You may need PPE (such as hearing protectors) because of plant working near you.

?

List the common points in the daily pre-use check.
What actions should you take when leaving mobile plant overnight near a public highway?
How, if at all, would you expect traffic to be controlled where a haul road crosses a site access road?

Encourage a discussion by using a real-life situation or example and ask if there are any questions.

# Lifting operations

| | |
|---|---|
| **Reason** | Lifting must be carried out in a safe manner. |
| **Why** | Unsafe lifting practices result in many accidents and injuries, including some fatalities each year. |
| **Outline** | This talk covers the procedures to follow to ensure that lifting operations are carried out in a safe manner. |

*Lifting equipment must be appropriate for the load*

# Lifting operations

| | |
|---|---|
| **Reason** | Lifting must be carried out in a safe manner. |
| **Why** | Unsafe lifting practices result in accidents and injuries, including some fatalities each year. |
| **Outline** | This talk covers the procedures to follow to ensure that lifting operations are carried out in a safe manner. |

## Types of lifting equipment

1. Lifting equipment includes items of plant, such as forklift trucks and telescopic handlers, all mobile elevating work platforms and cranes, electric hoists, goods hoists, gin wheels, and so on.
2. Slings, lifting beams, eye-bolts, and so on, are lifting accessories.

## Precautions

1. Risk assessments must be prepared for all lifting operations.
2. All lifting equipment and lifting accessories must be marked with their safe working load (SWL).
3. Lifting equipment and lifting accessories must not be used to move loads heavier than the marked SWL.
4. Lifting equipment must only be used by people who have been trained to do so.
5. Loads must only be slung by qualified persons. Never stand under a suspended load.
6. If using lifting equipment, look for overhead obstructions (such as power cables).
7. Ensure that lifting equipment has no obvious defects before using it.

## Forklift trucks and telescopic handlers

1. Travel with the load in the lowest position and don't raise it whilst travelling.
2. Ensure the load is stable and secure.
3. Do not carry passengers unless a passenger seat is fitted.
4. Do not use a forklift truck or telescopic handler to lift people unless it is suitably adapted.

## Cranes

1. A qualified appointed person must supervise all lifting operations.
2. It may be necessary to attach tag lines to the load to stabilise it when lifted.
3. Beware of changing weather conditions or wind speed making lifting operations unsafe.

## Mobile elevating work platforms

1. Use only on firm, level ground and use outriggers or stabilisers where necessary.
2. Users must always use a restraint lanyard clipped to the anchorage point on the machine.

What restriction is there on a load to be lifted by lifting equipment?
Who is allowed to operate items of lifting equipment?
How should the load be carried when travelling up a slope?
What checks should you carry out before use?

Encourage a discussion by using a real-life situation or example and ask if there are any questions.

# Lifting equipment and accessories

| | |
|---|---|
| **Reason** | Lifting accessories must be correct for the job, well maintained and properly used if lifting operations are to be carried out safely. |
| **Why** | Accidents have occurred because of the failure of lifting accessories. |
| **Outline** | This talk covers slings, chains, shackles and hooks and eyebolts. |

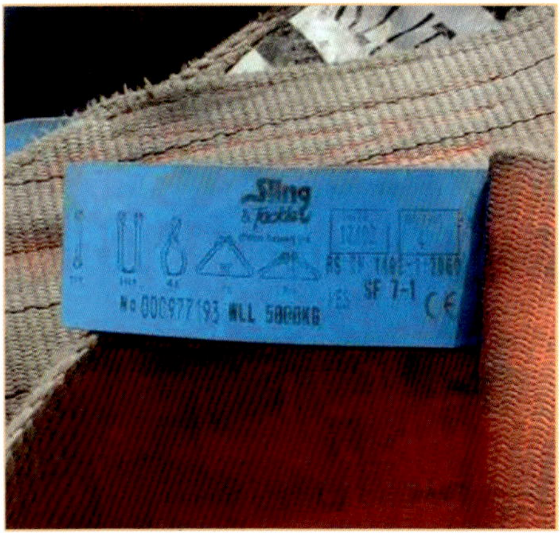


*Check condition and safe working load of strops*

*Check ID, safe working load and condition of chains*

*Correct slinging of a palletised load*

*Inspecting lifting equipment*

A chain is only as strong as its weakest link – know the SWL of all lifting gear.

# Lifting equipment and accessories

| | |
|---|---|
| **Reason** | Lifting accessories must be correct for the job, well maintained and properly used if lifting operations are to be carried out safely. |
| **Why** | Accidents have occurred because of the failure of lifting accessories. |
| **Outline** | This talk covers slings, chains, shackles and hooks and eyebolts. |

## Slings

1. Check the SWL on the sling against the weight of the load to be lifted.
2. Don't use fibre rope or wire slings for hot loads and keep them away from welding or flame-cutting operations.
3. Ensure there are no broken ends in wires and no chafing on fibre ropes.
4. Check the condition of splices, rings and thimbles on slings.
5. Remember – the load on each leg of a multi-leg sling increases as the angle between the legs gets wider.
6. Protect wire rope or nylon slings from sharp edges.

## Chains

1. Make sure that the chain is not kinked or twisted and don't shorten a chain by knotting it.
2. Never lengthen a chain by joining pieces together.
3. Don't lubricate chain slings because oil can pick up abrasive materials (such as sand or grit).
4. Don't expose chains to acids or corrosive substances.

## Shackles

1. Use the right type of shackle for the job in hand. Don't use any shackle that isn't marked with the SWL.
2. Check the bow and pin for damage and destroy it if there is any doubt.
3. When using a shackle with a nut and bolt pin, the pin should be free to rotate when the nut is tight.

## Hooks and eyebolts

1. Check both hooks and eyebolts carefully for cracks, cuts, dents and corrosion pits.
2. Swivel hooks should rotate freely.
3. Always mouse hooks unless fitted with a safety catch and make sure the catch operates freely.
4. Check the centre line of the eye is central with the threaded portion.

What must you check before using a sling?
What happens if you double a sling around a shackle?
What should you check on a shackle before use?
What checks should be carried out on hooks and eyebolts before using them?
Name two things you must not do when using chains.
If a hook isn't fitted with a catch, what should you do?

Encourage a discussion by using a real-life situation or example and ask if there are any questions.

# Hoists and hoist towers

| | |
|---|---|
| **Reason** | All persons must be aware of the safe methods of using hoists. |
| **Why** | A hoist can be a dangerous piece of equipment if used unsafely. |
| **Outline** | This talk covers the safe working practices when using hoists. |

*Large goods hoist*

*Small goods hoist*

| Notes |
|---|
| |
| |
| |
| |
| |
| |
| |
| |

# Hoists and hoist towers

| | |
|---|---|
| **Reason** | All persons must be aware of the safe methods of using hoists. |
| **Why** | A hoist can be a dangerous piece of equipment if used unsafely. |
| **Outline** | This talk covers the safe working practices when using hoists. |

## Construction and maintenance

1. The erection, alteration or dismantling of a hoist is a specialist operation and should only be carried out by a competent person.
2. Any fixed hoist tower must be adequately tied in to the host structure.
3. Hoists should be constructed in such a way to prevent the fall of materials from the platform or cage.

## Safety features

1. All hoist cages and platforms must be marked with the SWL.
2. All hoists must be marked whether they are for goods or passenger use.
3. Passenger hoists must be fitted with interlocked gates at every landing place.
4. All hoists should be fitted with an efficient braking device capable of supporting the platform and load in the event of a failure of the lifting gear.

## Operation

1. Hoists should only be operated by an authorised person.
2. All ground level and landing gates must be kept closed whilst the hoist platform or cage is in motion.
3. Passengers must never attempt to travel in hoists designed for goods only.
4. Hoists should never be operated in excess of the SWL.

## Inspection

1. All hoists must be subjected to periodic examination by a competent person.
2. Passenger lifts must be examined to test safety devices every time the height of the hoist way is altered.

What could be the consequence of interfering with hoist ties?
What action should you take if you believe that a hoist is defective?
What could be the consequence of overloading a passenger hoist?
Why should passengers not use a hoist intended for goods only?
Who should carry out the thorough inspection of a hoist?

Encourage a discussion by using a real-life situation or example and ask if there are any questions.

# Signallers and slingers

| | |
|---|---|
| **Reason** | Heavy objects are lifted around sites by cranes regularly – the potential for danger is obvious. |
| **Why** | Signallers and slingers must be trained and competent to sling and signal safely. |
| **Outline** | This talk covers lifting gear, before lifting, during lifting and potential hazards. |

*Slingers and signallers must have appropriate training*

You are the eyes of the crane driver.

# Signallers and slingers

| | |
|---|---|
| **Reason** | Heavy objects are lifted around sites by cranes regularly – the potential for danger is obvious. |
| **Why** | Signallers and slingers must be trained and competent to sling and signal safely. |
| **Outline** | This talk covers lifting gear, before lifting, during lifting and potential hazards. |

## Lifting gear

1. Check the lifting gear for kinks and frays each day. Chains must not be joined by means of bolts or wire.
2. Lifting gear must not be used unless its SWL is marked or otherwise known.
3. Don't use improvised slings or a single leg of a multiple sling.
4. Store chains, ropes, strops and slings in dry conditions.

## Before lifting

1. Wear appropriate PPE, for example a safety helmet and high-visibility clothing, because the machine operator will need to see you at all times, and you need to protect yourself from injury.
2. Make sure you know the weight of the load to be lifted.
3. Ensure hooks are 'C' type or fitted with a safety catch.
4. Ensure you can see the crane driver. If you can't, use radios that are fully charged at the start of the shift.

## During lifting operations

1. Use approved hand signals clearly and distinctly.
2. Protect wire ropes and slings from sharp edges of the load with softwood or other suitable packing.
3. Ensure the correct pin in the shackle is used and screwed home.
4. Ensure the hook is central to stop the load being dragged and/or swinging when raised.
5. Ensure the load is lifted off the ground, is free and is correctly slung before hoisting.
6. Always use a guide rope to steady the load and stand well clear of the load being lifted.
7. Warn the crane operator of any obstructions to the load.
8. To avoid damage to slings, loads should be landed on to timber baulks or other suitable bearers.

## Hazards

1. Never tie knots in chains to shorten them and don't use lifting gear for other purposes (such as towing).
2. Keep all those not involved in lifting operations away from the vicinity, especially children and the general public.

?

What should you check before lifting?
How do you protect lifting gear from sharp edges?
What must you look out for during lifting operations?

Encourage a discussion by using a real-life situation or example and ask if there are any questions.

# D

# High risk activities

## Working at height



## Excavations and buried services



## Confined space working

# Working at height

| | |
|---|---|
| **Reason** | Around 40% of major injuries on construction sites involve falls from heights. |
| **Why** | More than 50% of falls from height end in death. Don't end up as a statistic. |
| **Outline** | This talk covers what you need to think about and do, before and during working at height. |

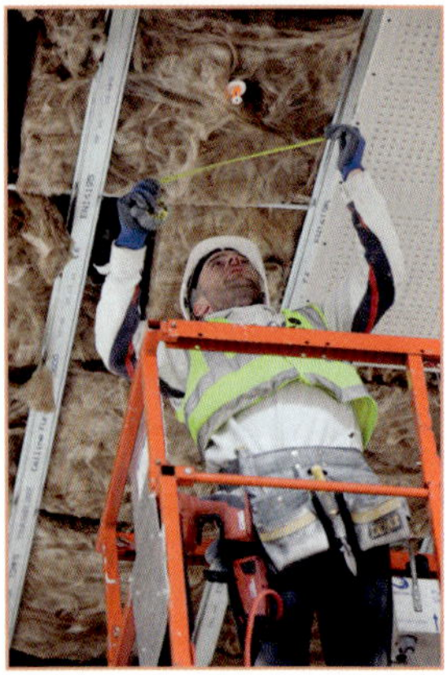

*Examples working at height activities with suitable control measures*

There is no safe height to fall from.

# Working at height

| | |
|---|---|
| **Reason** | Around 40% of major injuries on construction sites involve falls from heights. |
| **Why** | More than 50% of falls from height end in death. Don't end up as a statistic. |
| **Outline** | This talk covers what you need to think about and do, before and during working at height. |

## Before working at height

1. Ladders and stepladders are only for light work of a short duration and where there is a low risk of falling.
2. Ideally other, safer access equipment will be provided.
3. Roof-edge barriers (or scaffolds) must be erected therefore, preventing the risk of people and materials falling from height.
4. Access ladders must extend at least one metre (5 rungs) above the stepping-off point and must be secured.
5. Fragile surfaces (such as asbestos cement roof sheets and some skylights) must be identified with signs and measures taken to stop you falling through.
6. Ladders must be rested at the correct angle (one unit out to four units up).

## Hazards

1. Any fall could result in serious injury, so guard-rails and toe-boards or suitable access equipment must be installed and/or used.
2. Adverse weather conditions must be anticipated and suitable precautions taken: beware of wet, windy or icy weather.
3. Too much material on a working platform can make access difficult or make the scaffold unstable.
4. Don't allow rubbish to accumulate. Use a chute or lower materials properly.

## Safe working on roofs

1. Only competent operatives may be used for roofing works.
2. A safe system of work must be devised and implemented where the roof would otherwise be liable to collapse under a person's weight: usually a temporary platform is used.
3. If edge protection or a soft-landing system cannot be used, it may be necessary to use a harness and lanyard. A secure anchorage point and training will be necessary.
4. Bitumen boilers require a drip tray and fire extinguisher.
5. Openings must be covered or guarded; if removed for any reason, replace as soon as practical.

What should be identified before work at height starts?
Above what height must edge protection, toe-boards and guard-rails be erected?
When might a temporary platform be used?
How far should ladders extend above a stepping-off point?

Now inform the workforce of company policy on working at height, particularly the use of ladders. Encourage a discussion by using a real-life situation or example and ask if there are any questions.

# Fragile roofs

| | |
|---|---|
| **Reason** | On average, 7 people are killed each year after falling through a fragile roof or fragile roof light. Many others suffer a permanent disabling injury. |
| **Why** | Fragile roof incidents are not inevitable. They can be prevented by careful planning, using trained and experienced workers with suitable equipment and employing a high level of supervision. |
| **Outline** | This talk will cover some of the types of fragile roofs, what you need to do before and during the work and the requirements for working on a fragile roof. |

*Rooflights and glass form fragile roof areas*

*Proper planning and supervision reduce the likelihood of falls*

*It's not just old buildings that have fragile roof areas*

# Fragile roofs

| | |
|---|---|
| **Reason** | On average, 7 people are killed each year after falling through a fragile roof or fragile roof light. Many others suffer a permanent disabling injury. |
| **Why** | Fragile roof incidents are not inevitable. They can be prevented by careful planning, using trained and experienced workers with suitable equipment and employing a high level of supervision. |
| **Outline** | This talk will cover some types of fragile roof, what you need to do before and during the work, and the requirements for working on a fragile roof. |

## Types of fragile (non-load bearing) roofs

1. Cement sheeting – non-reinforced sheets, irrespective of profile type, for example asbestos.
2. Roof lights – these are difficult to see in certain light conditions and can be hidden when painted or covered in plants like moss.
3. Liner panels on built-up sheet roofing.
4. Corroded metal sheets.
5. Glass, including wired glass.
6. Rotten chipboard, or similar materials.
7. Other materials, including wood-wool slabs, slates and tiles.

## Before and during the work

1. Planning and supervision are key to reducing the risk of a fall. Work should not start until the findings of a risk assessment have been taken into account and control measures are in place.
2. First ask yourself if the work can be done in a safer way, for example from under a work platform.
3. If not, put a planned safe system of work in place before the work starts.
4. Ensure the planned safe system of work is used by everyone involved.
5. As work progresses, check that safe systems are still suitable. If not, stop work until they are adapted.
6. If nets, air bags or other soft landing systems are used, they should be properly installed and inspected regularly.

## Requirements for working on fragile roofs

1. Ideally a safe system of work should be identified to enable the job to be carried out without anyone having to go on the roof.
2. Where this is not possible, install perimeter edge protection and spread the load with a stable, temporary working platform with guard-rails.
3. Ensure all the work and access platforms are fitted with guard-rails. If this is not possible, install safety nets, air bags and so on underneath the roof or use a harness system.
4. Where harnesses are used, make sure they have adequate anchorage points and they are properly used.
5. Load-bearing surfaces that cannot be dislodged should be fitted over skylights, particularly where fragile skylights are fitted to an otherwise load-bearing roof.

?

Why is it important to be able to recognise a fragile roof?
What should your first thought be when thinking about working on a fragile roof?
Name three things that suggest the roof is fragile.

Now inform your workforce of the company policy and requirements for working on fragile roofs. Encourage a discussion by using a real-life situation or example and ask if there are any questions

# Working on scaffolds

| | |
|---|---|
| **Reason** | Falls from height account for over 50% of deaths in the construction industry. |
| **Why** | If you don't follow the guidance you may end up as a statistic or, at best, in hospital. |
| **Outline** | This talk covers access, loading, hazards, inspection and security. |

Scaffolding provides you with a safe working platform – don't abuse it.

# Working on scaffolds

| | |
|---|---|
| **Reason** | Falls from height account for over 50% of deaths in the construction industry. |
| **Why** | If you don't follow the guidance you may end up as a statistic or, at best, in hospital. |
| **Outline** | This talk covers access, loading, hazards, inspection and security. |

## Access

1. Don't climb up or down scaffolding tubes – use ladders or stairs provided.
2. Make sure the ladder is at the correct angle (one unit out to four units up or 75%).
3. Ensure ladders are tied in at both stiles, not the rungs, and extend a safe distance (one metre) above the landing stage.
4. At the end of work, remove access ladders or board them up to prevent children playing on them.

## Loading

1. Don't overload scaffolding, position heavy loads adjacent to the standards as they are the load-bearing members, not in the centre of bays.
2. When stacking materials, always leave a passageway at least two boards wide for other people to pass, or three boards if wheelbarrows are in use.
3. Ensure materials are stacked correctly and can't fall. Use brick guards where necessary.
4. Don't leave tools or materials lying about on the platform.

## Hazards

1. Guard-rails and toe-boards must be fitted where a person is liable to fall and be injured.
2. Don't use incomplete scaffolding.
3. Don't remove or interfere with ties, guard-rails, bracing, toe-boards and ladders. Alterations must only be made by competent persons.
4. Don't throw, drop or tip materials from heights. Either lower or dispose of them through a chute.
5. The gap between toe-boards and mid guard-rails, and between mid and top guard-rails, must not exceed 470 mm.

## Inspection

1. Carry out a quick visual check at the start of each shift before going on to the scaffold.
2. Report any suspected faults or defects immediately.
3. Scaffolding should be inspected every seven days by a competent person and details of the inspection recorded.

?

At what angle should the access ladder be?
What action should you take on noticing defective scaffolding?
How should you dispose of material from heights?
What two things must you consider when loading scaffolding?

Encourage a discussion by using a real-life situation or example and ask if there are any questions.

# Mobile towers and access equipment

| | |
|---|---|
| **Reason** | Mobile towers must be erected in accordance with the manufacturer's instructions and only by competent persons. |
| **Why** | Unless you use a mobile tower correctly you could find yourself seriously injured or even worse. |
| **Outline** | This talk covers erection, use, stability and hazards. |

*Safe working on a mobile tower, with toe-boards and both guard-rails in position and wheels locked*

*A proprietary access system incorporating ladder access, working platform, guard-rails and hoist*

| Notes |
|---|
| |
| |
| |
| |
| |
| |

After moving a tower and before use the wheels must be locked.

# Mobile towers and access equipment

| | |
|---|---|
| **Reason** | Mobile towers must be erected in accordance with the manufacturer's instructions and only by competent persons. |
| **Why** | Unless you use a mobile tower correctly you could find yourself seriously injured or even worse. |
| **Outline** | This talk covers erection, use, stability and hazards. |

## Before erecting the tower

1. Ensure you are competent to erect the tower.
2. Choose a safe method of work as recommended by the Prefabricated Access Suppliers' and Manufacturers' Association (PASMA).
3. Check all components are in a good condition.
4. Check wheels for effective rotation.
5. Check brakes and locking devices work correctly.

## Before use

1. Ensure the tower is vertical and square.
2. The wheels must be locked when the tower is not being moved.
3. Check that outriggers are set correctly and secured if needed.
4. Ensure the platform is fully boarded out and guard-rails and toe-boards are fitted.
5. Be aware of weather conditions if the tower is outdoors: high winds can blow them over.

## Access to the platform

1. Always use the built-in stairway or ladder. If it is a vertical ladder, always climb it on the inside of the tower.
2. Follow the manufacturer's instructions on base to height ratio.
3. Guard-rails should be at least 950 mm high, and an intermediate guard-rail should be provided so that the unprotected gap does not exceed 470 mm.
4. Tie the tower to a permanent structure where necessary.
5. Don't move the tower if people or materials are still on the platform. Don't pull the tower along while standing on it.

## Safe use

1. Don't exceed the manufacturer's safe working load (SWL) for the tower.
2. When moving towers, ensure there are no potholes, obstructions or overhead power lines in the way.
3. When working, ensure the access hatch is closed on the platform.
4. Never use ladders or steps on a scaffold platform to get extra height.
5. Towers must only be used on firm surfaces. Provide support on soft ground.
6. Do not remove components (such as guard-rails) if they are stopping you doing your job.
7. Always install stabilisers or outriggers when advised to do so in the instruction manual.

When should toe-boards and guard-rails be fitted?
How should you hoist materials on to a working platform?
What hazards must be considered when moving a tower?
Outline one pre-erection and one pre-use check.
Outline one stability consideration and one hazard.

Encourage a discussion by using a real-life situation or example and ask if there are any questions.

# Low-level access equipment (trestles, podiums and hop ups)

| | |
|---|---|
| **Reason** | Low-level access equipment is often misused, which can result in accidents. |
| **Why** | Being aware of the rules and following them can help to prevent a fall. |
| **Outline** | This talk covers some different types of low-level access equipment and how to use them safely. |

*Trestle systems are safer options*

*Hop ups should be used with care*

If you have to choose between a hop up or podium steps, remember that a handrail will stop you stepping off the edge – podiums are safer than hop ups.

# Low-level access equipment (trestles, podiums and hop ups)

| | |
|---|---|
| **Reason** | Low-level access equipment is often misused, which can result in accidents. |
| **Why** | Being aware of the rules and following them can help to prevent a fall. |
| **Outline** | This talk covers some different types of low-level access equipment and how to use them safely. |

## Folding and adjustable trestles

1. Guard-rails cannot generally be fitted to folding and adjustable trestles and so these should be avoided, if possible.
2. If used, they should only be used for light work and work of short duration.

## Modern trestle systems

1. Modern trestle systems should be safe and stable in use, with safe ladder access, guard-rails and toe-boards fitted.
2. They are similar in usage and appearance to a low-level scaffold platform.
3. A tied ladder must be available for access to the trestle.
4. Be aware of any weight restrictions before loading the trestle with materials.

## Podiums and hop ups

1. Podium steps are commonly accepted as the replacement for stepladders and are often overrated in their capability.
2. Training is important to understand their limitations, particularly with overreaching, locking wheels and use of stabilisers, if fitted.
3. Hop ups are small, flat platforms, which provide extra height (up to 600 mm).
4. Many construction sites do not allow them to be used as there is no provision for the prevention of falls.
5. Where hop ups are used great care must be taken to ensure the ground surface is stable.

**!** All work at height should be planned, organised and carried out by competent persons.

**?** What is the support spacing when using scaffold boards?
Why are many trestle systems unsafe to use?
Why shouldn't you use boards on step treads?
What safety features should you be looking for before using a modern type trestle system?
Why are podium steps only considered as a stepladder substitute?

Now inform your workforce of the company policy on the use of trestles, podiums and hop ups. Encourage a discussion by using a real-life situation or example and ask if there are any questions.

# Ladders and stepladders

| | |
|---|---|
| **Reason** | Ladders are probably the most used and misused pieces of access equipment. |
| **Why** | Using ladders safely will help to prevent accidents. |
| **Outline** | This talk covers pre-use checks, use of ladders, the associated hazards and restrictions. |

*Attach ladders securely*

What checks should you carry out before use?
At what angle should a ladder be placed?
How can the stability of the ladder be improved?
What precautions should you take when using a metal ladder?
What should you do on finding a defective ladder?
What is the correct way to climb a ladder?
Before using a stepladder, what should you check?

Now inform your workforce of the company policy regarding the use of ladders.
Encourage a discussion by using a real-life situation or example and ask if there are any questions.

# Ladders and stepladders

| | |
|---|---|
| **Reason** | Ladders are probably the most used and misused pieces of access equipment. |
| **Why** | Using ladders safely will help to prevent accidents. |
| **Outline** | This talk covers pre-use checks, use of ladders, the associated hazards and restrictions. |

## Pre-use checks

1. Ladders must be stored correctly and inspected regularly.
2. Check for splits or cracks in the stiles and rungs and that none of the rungs are missing or loose.
3. Don't use painted ladders as paint can hide damaged parts.
4. Report defects, label as defective and remove from site.

## Safe use of ladders

1. Only light work of a short duration, with a low risk of falling, should be carried out from a ladder.
2. Ladders should be set on a firm base and lean at the correct angle (one unit out to four units up).
3. Ladders must be tied near the top and extend a safe distance above the landing stage, unless a separate handhold is provided (usually one metre or five rungs).
4. If it can't be secured at the top, it may be possible to secure it at the bottom.
5. Ensure your footwear is free from excessive mud or grease before you climb up the ladder.
6. When climbing up or down, use both hands on the stiles. Always face the ladder.
7. Don't overreach from a ladder. Always move it.
8. If a task requires a worker to carry more than 10 kg (for example, a full bucket) up the ladder or steps it will need to be justified by a detailed manual handling assessment.
9. On a stepladder where a handhold is not practicable a risk assessment must identify whether it is safe to use.

## Hazards

1. Never stand a ladder on a drum, box or other unstable base.
2. Never attempt to repair broken ladders.
3. Never carry loads up ladders – use a hoist.
4. Ladder rungs must not be used as improvised ramps.
5. When using metal or metal-reinforced ladders, make sure there are no electrical hazards nearby.

## Restrictions

1. The law requires that careful consideration is given before the use of a ladder is approved.

## Stepladders

1. Use a mobile scaffold tower or MEWP, if possible, instead.
2. Check treads, stiles, hinges and restraining rope before use.
3. If the stepladder is damaged tell your supervisor and place it where no-one else can use it.
4. Only ever use on a firm, level base and don't work higher than two-thirds up the stepladder (handhold).
5. Do not place boards between pairs of stepladders to form an improvised working platform.
6. Wooden steps should not be painted; it might hide defects.
7. Don't over-reach or apply a sideways loading to stepladders – return to ground level and move them.
8. Ensure steps are fully extended before you go up.

# Mobile elevating work platforms

| | |
|---|---|
| **Reason** | MEWPs are used extensively to gain access on construction sites. |
| **Why** | They are useful items of plant if used correctly, but dangerous if not used in a safe manner. |
| **Outline** | This talk covers the hazards and safe operating methods for MEWPs. |

*MEWP in use*

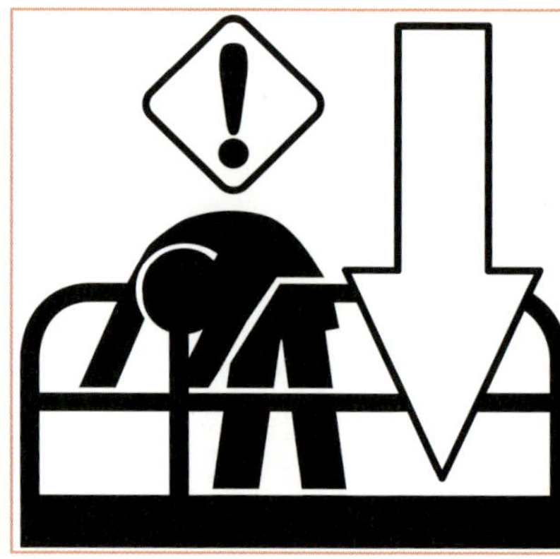

*MEWP emergency descent symbol*

*MEWP operators should be trained*

# Mobile elevating work platforms

| | |
|---|---|
| **Reason** | MEWPs are used extensively to gain access on construction sites. |
| **Why** | They are useful items of plant if used correctly, but dangerous if not used in a safe manner. |
| **Outline** | This talk covers the hazards and safe operating methods for MEWPs. |

## Hazards

1. Operatives falling from height due to unsafe work practices.
2. Overturning of the machine due to poor operating technique or unsatisfactory ground conditions.
3. Collision with other vehicles (knuckle or elbow of boom moving into the path of other traffic).
4. Tools and materials falling from height.
5. Contact with high level, live electrical cables and other obstructions.
6. Exhaust fumes, if used in a confined area.
7. High wind speeds and other adverse weather conditions.

## Precautions

1. All operators of MEWPs must be trained in their use.
2. Operators should only operate the types of MEWP for which they have been trained.
3. Always check that the machine is stable before use.
4. Use outriggers, or stabilisers, where necessary.
5. Generally, for all MEWPs, except scissor lifts, users should use a restraint lanyard clipped to the correct anchorage point on the machine.
6. For scissor lifts, the need to use a safety harness or lanyard will depend upon the risk assessment.
7. Never clip onto the adjacent structure.
8. Ensure ground conditions are suitable for the type of machine; voids and drain covers can collapse.
9. Do not load the machine beyond its SWL.
10. If your work involves removing equipment or materials from a structure, don't forget to allow for the extra loading.
11. When manoeuvring in a confined area or where members of the public are at risk, always use a signaller.
12. Be prepared to stop work and return to ground level if the wind speed or weather conditions deteriorate to an unacceptable level.

## Refuelling

1. Always turn the engine off before refuelling.
2. LPG-powered machines must be refuelled in open spaces where any spillage can quickly disperse.
3. It is good practice to carry out refuelling of all machines in the open air.
4. Avoid skin contact if refuelling diesel oil, and clean up any spillage to avoid a slipping hazard.

What particular hazards are present in your current work area?
Why do you not clip onto the structure?
How can you find out who is qualified to operate a MEWP on site?
Where is spare fuel stored and refuelling carried out on this site?

Encourage a discussion by using a real-life situation or example and ask if there are any questions.

# Fall arrest and suspension equipment

| | |
|---|---|
| **Reason** | Serious injuries on site caused by falls, falling materials and equipment need to be reduced. |
| **Why** | 50% of people who fall from heights in excess of over two metres die. Don't become a fatality. |
| **Outline** | This talk covers aspects that need consideration before, during and after using soft-landing systems and suspension equipment. |

*Different types of safe working at height*

*Operative using a fall arrest block to reduce the risk of falling where no collective edge protection is available*

*Attach lanyards securely to the correct anchor point*

# Fall arrest and suspension equipment

| | |
|---|---|
| **Reason** | Serious injuries on site caused by falls, falling materials and equipment need to be reduced. |
| **Why** | 50% of people who fall from heights in excess of over two metres die. Don't become a fatality. |
| **Outline** | This talk covers aspects that need consideration before, during and after using soft-landing systems and suspension equipment. |

## Before use

1. Only trained and competent persons must undertake the installation of this type of equipment.
2. The equipment should be inspected to check that it is serviceable and suitable for the current task.
3. Where safety nets will be used, all the anchorage points and supports must be suitable and secure.
4. Airbags must be inflated with the pump left running.
5. Work must stop if the prevailing weather conditions make it unsafe to continue.
6. Users of harness and lanyards must be trained in their inspection and use.

## Safe use

1. Only authorised, trained and competent personnel are allowed to use suspension equipment.
2. Regular checks must be made to ensure that safety nets remain free from rubbish and debris.
3. Deliberate interference with any safety-critical equipment is a criminal act.
4. The immediate area must be free from projections that could impair the safe use of any suspended cradle.
5. The area below where work is being carried out at height might have to be designated as a danger area.
6. The free end of each safety lanyard must be clipped onto a designated anchorage point.
7. All tools must be adequately secured to prevent them being dropped to the area below.

## After use

1. All equipment is to be inspected for damage and must be clean and dry before being stowed away.
2. Any defects with the equipment are to be reported promptly and correctly.
3. Adequate records of the condition and usage of the equipment should be maintained.
4. All tie-offs for cradles, ancillary lines and ropes are to be left in a secure position to prevent unauthorised access and use of the equipment.

What should happen to safety nets and suspension equipment after use?
Who should be informed that the equipment is to be used?
What should happen to the area below the equipment?

Now inform your workforce of the company policy regarding the use of fall arrest and suspension equipment.
Encourage a discussion by using a real-life situation or example and ask if there are any questions.

# Working over safety nets or soft-landing systems

| | |
|---|---|
| **Reason** | Falls from height account for 50% of all accidents and deaths within the construction industry. |
| **Why** | Using safety nets and soft-landing systems safely can save lives or minimise injury in the event of a fall. |
| **Outline** | This talk covers the use of safety nets and soft-landing systems, together with some requirements for installation and rescue. |

*Safety nets installed prior to roof works commencing*

| Notes |
|---|
| |
| |
| |
| |
| |
| |
| |
| |

# Working over safety nets or soft-landing systems

| | |
|---|---|
| **Reason** | Falls from height account for 50% of all accidents and deaths within the construction industry. |
| **Why** | Using safety nets and soft-landing systems safely can save lives or minimise injury in the event of a fall. |
| **Outline** | This talk covers the use of safety nets and soft-landing systems, together with some requirements for installation and rescue. |

## Safe use

1. There are many different types of safety nets and soft-landing systems that can be selected to suit particular circumstances. They are not designed to prevent falls, but to minimise the risk of injury after a fall, of either people or materials.
2. Safety nets should be fitted as high as possible beneath the work area to minimise the distance of a fall.
3. They must be tight enough to minimise sag, when loaded, and prevent the load hitting the ground or floor below.
4. They must be installed by qualified people and inspected weekly by a competent person.
5. Soft-landing systems are usually large bags that are filled with air, via a pump, or a group of smaller, pre-packed bags (beanbags), with polystyrene chippings, clipped together by plastic clips. The depth of the bag reduces the distance of the fall and also cushions the fall.
6. These systems are known as collective fall-protection measures and are preferred to the use of safety harnesses and lanyards.

## Means of rescue

1. When safety nets are used, consider how someone will get out of the net if a fall takes place.
2. A means of rescue must be planned and in place from the outset.
3. If a person falls and is not injured they can climb out, but if they are incapacitated they will need to be rescued, preferably from below the net to prevent unnecessary disturbance to the injured person by rescuers climbing into the net.
4. Soft-landing systems have similar requirements in terms of recovery from a fall but in this case rescue from below is not possible.
5. When using airbags it is possible to slowly reduce the air pressure to gain access that will not disturb a casualty too much. With beanbags, which are a little more rigid, boarding could be used if readily available or the bags could be unclipped and removed to a walkway.

## Inspection

1. Safety nets are individually tagged (identified), have to be inspected weekly (as well as daily pre-user checks) and the inspection has to be recorded.
2. Soft-landing systems should be subjected to daily pre-user checks but these do not have a statutory inspection requirement.
3. Any safety net or soft-landing system that has been subjected to a load (a fall of a person or materials) may have been deformed or unplaced and must be examined before anyone is allowed to work above that area again.
4. Sharp objects are also likely to cause cuts, which would weaken a net or cause deflation of an airbag.

When should you check a safety net or soft-landing system?
When should a safety net be examined?
What is the best way to rescue an injured person from a safety net?

Now inform your workforce of the company provision for rescue from nets and safe-landing systems. Encourage a discussion by using a real-life situation or example and ask if there are any questions.

# Excavations

| | |
|---|---|
| **Reason** | People die in trenches. The vast majority of fatal accidents occur in trenches that are less than 1.5 metres deep. |
| **Why** | A cubic metre of earth can weigh over a tonne – the only body able to support that is a dead one. |
| **Outline** | This talk covers how you can stay safe in an excavation and outlines the causes of accidents. |

*Safe working in a deep excavation using trench boxes and a tied ladder*

| **Notes** |
|---|
| |
| |
| |
| |
| |

A cubic metre of earth can weigh 1.5 tonnes – if you get it wrong the consequences could be fatal.

# Excavations

| | |
|---|---|
| **Reason** | People die in trenches. The vast majority of fatal accidents occur in trenches that are less than 1.5 metres deep. |
| **Why** | A cubic metre of earth can weigh over a tonne – the only body able to support that is a dead one. |
| **Outline** | This talk covers how you can stay safe in an excavation and outlines the causes of accidents. |

## How to stay safe

1. Before digging check for services (water, gas and electric). Always treat them as live, until proven otherwise.
2. Excavations must be supported or battered back where necessary to prevent collapse.
3. Use ladders for access and egress, do not climb supports.
4. Fit edge protection around excavations to protect the general public, regardless of excavation depth.
5. Keep spoil heaps back at least the depth of the excavation from the edge.
6. Ensure stop blocks are fitted when a vehicle or plant is tipping into an excavation and that the vehicle or plant is guided by a signaller.
7. Wear your safety helmet at all times.
8. Never throw tools or materials into an excavation, pass them hand-to-hand or lower them on a rope if it is too deep to pass them.
9. Excavations must be inspected by a competent person prior to entry at the start of every shift.
10. The results of at least one inspection in every seven-day period must be recorded.

## Causes of excavation accidents

1. Shoring not installed or trench not battered back.
2. Workers trying to jump across the excavation.
3. Unauthorised removal or alteration of supports or braces.
4. People working beyond the supported area of the excavation.
5. Sides of excavations becoming unstable after periods of hot, dry weather or heavy rainfall.
6. Heavy materials, vehicles or plant being placed, driven or parked too close to the edge, causing it to break away.
7. Vehicles driving into excavations due to lack of stop-blocks, barriers or bunds.
8. People, including members of the public, falling into trenches because edge protection was not fitted.
9. Workers not using the ladder for access or egress.
10. The accumulation of heaver-than-air toxic gases.

What must you check for prior to excavating?
What precautions must be taken when dumpers are tipping spoil back into a trench?
What is the safe means of access to a deep excavation?
What should be fitted to stop vehicles driving into excavations?
When should edge protection be fitted?

Encourage a discussion by using a real-life situation or example and ask if there are any questions.

# Buried services

| | |
|---|---|
| **Reason** | Every year many people are injured and some killed due to contact with buried services. |
| **Why** | 70,000 instances of damage and 150 injuries per year show there is room for improvement. |
| **Outline** | This talk covers gas, water mains, sewers and colour-coding for buried services. |

*Shallow buried services*

*Types of marker posts commonly found to indicate the location of buried services*

*Avoid contact with any damaged cable or apparatus*

All buried services have the potential to cause injury.

# Buried services

| | |
|---|---|
| **Reason** | Every year many people are injured and some killed due to contact with buried services. |
| **Why** | 70,000 instances of damage and 150 injuries per year show there is room for improvement. |
| **Outline** | This talk covers gas, water mains, sewers and colour-coding for buried services. |

## Gas

1. Check the gas company plans before digging.
2. Dig carefully by hand to establish the location of pipes and mark the route of all known pipes.
3. Remember gas is flammable and explosive.
4. At the slightest hint of gas escape, leave the area and do not smoke. Call the gas company and emergency services. The gas emergency number is 0800 111 999.
5. Modern, smaller diameter house mains are often plastic. Don't confuse them with electric cables.
6. Follow the gas company specifications for back-filling.

## Water mains

1. Trace the line of the main by trial pits and mark the route of all known pipes.
2. Burst pipes can fill an excavation quickly. If damaged, call the water company.
3. Remember, water at high pressure could be dangerous.
4. Don't leave a length of pipe unsupported.

## Sewers

1. There is a severe health risk if a foul sewer is fractured – leave the excavation and report it.
2. Wear personal protective equipment (PPE) due to the risk of contamination from sewage. Wash your hands before eating or smoking.
3. If you break a stormwater sewer and rain is falling, vacate the excavation as flooding may occur.

## Colour-coding system for buried services

1. **Black** or **red** – Electricity
2. **Blue** – Water
3. **Yellow** – Gas
4. **Grey** or **black** – Telecommunications
5. **Green** – Cable television

What must you do before digging?
If you notice gas escaping, what should you do?
How would you find the line of the water main?
What must you remember with unsupported pipes?
What hazards are there when working on sewers?
What are the colour-codings for buried services?
Before back-filling, what must you find out?

Encourage a discussion by using a real-life situation or example and ask if there are any questions.

# Confined spaces

| | |
|---|---|
| **Reason** | Working in confined spaces has led to the deaths of many workers. |
| **Why** | Through not knowing the associated dangers, workers or rescuers can become fatalities. |
| **Outline** | This talk covers hazards, before entry, work in confined spaces and emergency procedures. |

*Confined spaces training*

*Check the atmosphere before entering*

Don't enter a confined space until the risks have been assessed.

# Confined spaces

| | |
|---|---|
| **Reason** | Working in confined spaces has led to the deaths of many workers. |
| **Why** | Through not knowing the associated dangers, workers or rescuers can become fatalities. |
| **Outline** | This talk covers hazards, before entry, work in confined spaces and emergency procedures. |

## Hazards

1. Oxygen-depleted or enriched environments and suffocating, toxic or flammable atmospheres.
2. Actual or potentially hostile environments (inside plant).
3. Biological hazards (such as Weil's disease from rats' urine).
4. Confined spaces include cellars, chambers, pits, tanks, manholes, sewers, tunnels and some excavations.

## Before entering

1. Make sure a risk assessment has been carried out by a competent person.
2. Check for flammable or toxic gases and oxygen content.
3. If breathing apparatus is required, don't enter until you have been trained to use it.
4. Obey permits to work and check communications and monitoring equipment.

## Working in confined spaces

1. Work will be controlled by a permit to work, which will include arrangements for rescue. No rescue team – no entry!
2. Wear the protective equipment and clothing provided and continuously monitor the air quality.
3. Only BASEFFA-approved electrical equipment is to be used where flammable gases may be present.

## Emergency procedures

1. Make sure the recovery winch and apparatus is working.
2. Locate the position of the nearest telephone and understand the emergency procedure.
3. Don't attempt a rescue unless you are part of a trained rescue team.
4. The first duty of any rescuer is to ensure their own safety.
5. Leave a confined space immediately if told to do so.
6. REMEMBER to always follow the rescue plan.

?

What hazards can be found in confined spaces?
What must you check before entering a confined space?
Name three safety points regarding confined spaces.
As a rescuer, what must you be wearing?
What areas could be classed as confined spaces?
Who should carry out an assessment of the confined space before entry?

Now inform your workforce of the company policy regarding working in confined spaces.
Encourage a discussion by using a real-life situation or example and ask if there are any questions.

# E
# Environment

## Environmental management



## Waste management

# Being a good neighbour

| | |
|---|---|
| **Reason** | Many local communities regard construction work as a nuisance. |
| **Why** | If you understand people's concerns you can help to minimise the impact of your work. |
| **Outline** | This talk covers some of the problems and how you can help to reduce the impact. |

*Good relationships with the local community are vital*

| **Notes** |
|---|
| |
| |
| |
| |
| |
| |
| |

Always be polite and considerate. Take notice of any complaint levelled at you or the site and report it to your supervisor.

# Being a good neighbour

| | |
|---|---|
| **Reason** | Many local communities regard construction work as a nuisance. |
| **Why** | If you understand people's concerns you can help to minimise the impact of your work. |
| **Outline** | This talk covers some of the problems and how you can help to reduce the impact. |

## Being a good neighbour

1. It is possible that local residents may have objected to the site work commencing, and so they may feel aggrieved that it is now happening.
2. Everyone has a responsibility to work with the community and ensure good relations by minimising disruption and being considerate.
3. The workforce must have a positive influence on the local community, including residents, businesses and schools, and respect them by being a good neighbour, whilst at the same time avoiding complaints and possible prosecution.

## Control measures

1. Always be polite and considerate. Take notice of any complaint levelled at you or the site and report it to your supervisor.
2. Keep disruption from the site to a minimum by minimising dust, noise and vibration (from, for example, piling or breaking up solid materials).
3. Arrange for deliveries when traffic flow is likely to be low and avoid school arrival and departure times.
4. Talk to the local community and tell them what's happening. Ask for their concerns, if any.
5. Maintain good housekeeping by keeping roads and pathways clean, and minimising waste piles and overloaded skips that may present a fire or contamination risk.

## Precautions

1. Do not park vehicles so they cause an obstruction to driveways, footpaths or roads.
2. Do not trespass on neighbours' land.
3. Do not cause excessive noise by leaving engines running, shouting or having radios playing too loudly.
4. Do not whistle at passers-by or be inappropriately dressed.

?

What action should be taken when a complaint is received?
What are the activities on this site that are likely to cause the most nuisances?
How can we improve relations with our neighbours?

Now inform your workforce of the company policy on the company code of considerate practice for this site.
Encourage a discussion by using a real-life situation or example and ask if there are any questions.

# Environmental nuisance

| | |
|---|---|
| **Reason** | Where we work and what we do can cause an environmental nuisance. |
| **Why** | If you understand the problems caused by site activities you are in a better position to minimise the impact. |
| **Outline** | This talk covers some of the things that can cause an environmental nuisance. |

*Dust and noise can cause a nuisance beyond the construction site*

| **Notes** |
|---|
| |
| |
| |
| |
| |
| |
| |
| |

Your activities today may bring an early death to you and to others exposed to your inappropriate actions – so don't do anything that might cause an environmental or health problem.

# Environmental nuisance

| | |
|---|---|
| **Reason** | Where we work and what we do can cause an environmental nuisance. |
| **Why** | If you understand the problems caused by site activities you are in a better position to minimise the impact. |
| **Outline** | This talk covers some of the things that can cause an environmental nuisance. |

## Causes

1. Environmental nuisance is an activity that causes a disturbance.
2. The creation of dust, odour, smoke and other emissions may cause health risks to you or to other people near the site.
3. Dust can cause damage to vegetation and crops, the wildlife and watercourses.
4. Odours and smoke can cause breathing problems when inhaled and lung diseases resulting in a shortening of life.
5. Poor housekeeping will help create an environment that causes a nuisance.

## Control measures

1. Dampen down traffic routes and use wet cutting to reduce the potential for creating dust.
2. Minimise dropping heights from a bucket into a haulage vehicle to reduce noise and sheet the loads to reduce the potential for spills and dust.
3. Keep to site speed limits and using wheel-wash facilities to help reduce noise, ground vibration and dust.
4. Well planned and safe storage for materials prevents the additional potential for airborne nuisance and contamination.
5. Do not ignore complaints. Respond politely and inform your supervisor.
6. Most importantly, monitor activities and report any shortfalls to your supervisor.

## Precautions

1. Obtain permission from your supervisor or site manager before having a bonfire on site.
2. If plant is not properly maintained the engine can produce more toxic fumes.
3. Avoid leaving engines running when they are not in use.

?

What should you do if you receive a complaint from somebody?
What action should you take if your machine is overdue for a service and the exhaust smoke is black?
What should you do if you see a delivery vehicle driving too quickly?

Now inform your workforce of the company provision for reduction of environmental nuisance. Encourage a discussion by using a real-life situation or example and ask if there are any questions.

# Emergency spill control

| | |
|---|---|
| **Reason** | Spills on site cause damage to the environment and can harm animals, plants, fish and humans. |
| **Why** | If you know how to act in the event of a spill you can help protect the environment and save costly clean-ups. |
| **Outline** | This talk covers some of the methods used and the equipment involved to control spills. |

*Spill station*

*Spill kit in use*

It is important that everyone on site knows how to control a spill to minimise the impact. You should also know what equipment is available and where to find it.

# Emergency spill control

| | |
|---|---|
| **Reason** | Spills on site cause damage to the environment and can harm animals, plants, fish and humans. |
| **Why** | If you know how to act in the event of a spill you can help protect the environment and save costly clean-ups. |
| **Outline** | This talk covers some of the methods used and the equipment involved to control spills. |

## Emergency spill control

1. Accidental releases of fuels, oils and chemicals from construction sites make up a large number of the pollution incidents that happen each year.
2. Most spillages can be avoided with care and control.
3. It is very important that everyone on site knows how to control a spill, what equipment is available and where it is, so they can help to minimise the impact. They also need to know how to correctly dispose of spilled material.
4. Spill kits come in a variety of forms, including absorbent pads, socks, granules, pillows and wipes. Drain covers and barriers are also used.
5. An assessment of the potential areas of harm will be carried out on site and suitable kits will be placed in appropriate areas.

## Control measures

1. Know where the spill kits are on site. Make sure you know what to use and when, and what protective measures you need, including personal protective equipment (PPE).
2. Stop work to deal with a spill and tell your supervisor as soon as possible.
3. If the spill is likely to be flammable remove potential ignition sources.
4. Contain the spill with either a spill kit or use available materials to create a bund to prevent it from spreading.
5. Clean up manageable spills and place used absorbent material into bags for later safe disposal.

## Precautions

1. Do not put yourself in danger of exposure to harm from spills that you cannot identify. Seek advice.
2. Do not hose down or bury spills.
3. Do not allow spilt materials to enter a drain, gully or watercourse.
4. Do not store harmful materials within 10 metres of a watercourse.

When should you report a spillage and to whom?
Why is it important to prevent spillages?
What is the most important thing to remember with regard to personal safety?

Now inform your workforce of the company provision for emergency spill control.
Encourage a discussion by using a real-life situation or example and ask if there are any questions.

# Cement, concrete and plaster

| | |
|---|---|
| **Reason** | Some types of materials in construction are known to cause harm to the user when working with them unless precautions are taken. |
| **Why** | If you recognise the problems and know what action to take you can help to prevent environmental and personal harm. |
| **Outline** | This talk covers some of the types of materials involved and some controls. |

*Take precautions when working with concrete and cement*

Wet concrete is potentially hazardous on the skin as it burns and causes considerable damage to the cell structure. It has been the cause of amputations.

# Cement, concrete and plaster

| | |
|---|---|
| **Reason** | Some types of materials in construction are known to cause harm to the user when working with them unless precautions are taken. |
| **Why** | If you recognise the problems and know what action to take you can help to prevent environmental and personal harm. |
| **Outline** | This talk covers some of the types of materials involved and some controls. |

## Materials in their dry and wet state

1. All dusty materials (cement, lime, plaster and even sand) can cause health hazards to the skin, eyes and lungs.
2. Cement, concrete and plaster should be disposed of and washed out only at the designated locations.
3. Only use designated washout areas to clean off plant and equipment, as the dirty water will pollute watercourses if not disposed of properly.
4. Wet concrete and mortar are potentially hazardous on the skin as they burn and cause damage to the cell structure. Where incorrect or inadequate PPE has been worn, in some cases this has resulted in the need for amputations.
5. Poor storage facilities, damp environments, burst bags or leaking bulk storage systems cause potential environmental hazards, particularly to ground contamination and watercourses.

## Control measures

1. Be aware of personal hygiene, wear the correct PPE and be aware of the need for replacements.
2. Ensure that the storage area of materials is suitable.
3. Carry out mixing and batching works in areas well away from watercourses, gullies and drains.
4. Use designated wash out areas and ensure that delivery drivers (of concrete or similar) are aware of where they can wash out.
5. Minimise your exposure to dust from dry materials by wearing the correct PPE.

## Precautions

1. Do not hose down spills of materials into watercourses or drains.
2. Do not allow wash out water into watercourses or drains.
3. Do not wear inappropriate PPE – your health is important to both you and your employer.

What should you do if you do not have the correct PPE for your work?
Where is the right place to wash out a container used for mixing rendering plaster?
Why is it important not to let discharges into watercourses or drains?

Now inform your workforce of the company provision for delivery, handling, storage and safe use of materials.
Encourage a discussion by using a real-life situation or example and ask if there are any questions.

# Pumping, over-pumping and washing down plant

| | |
|---|---|
| **Reason** | Construction activities often create water contaminated with content harmful to the environment. |
| **Why** | If you know how to contain and manage contaminated water you will avoid causing pollution. |
| **Outline** | This talk covers the types of activity and some control measures. |

*Using a silt trap*

*Contain contaminated water*

If you know how to contain and manage contaminated water you will avoid causing pollution and damage to the environment.

# Pumping, over-pumping and washing down plant

| | |
|---|---|
| **Reason** | Construction activities often create water contaminated with content harmful to the environment. |
| **Why** | If you know how to contain and manage contaminated water you will avoid causing pollution. |
| **Outline** | This talk covers the types of activity and some control measures. |

## Contaminated water

1. Excavations often require dewatering, such as the removal of ground or rainwater, which may contain silt and other contaminants (for example, previously developed brownfield land).
2. Silt is composed of very fine particles of soil that, when mixed with water, creates mud that can be washed off sites into nearby watercourses causing harm to wildlife and humans.
3. Bentonite is a type of clay that swells and gels when dispersed in water. The use of Bentonite can lead to spillage around operational and mixing areas. Bentonite, in a liquid form, is highly polluting if it enters watercourses.
4. Water from washing down plant and machinery is likely to contain not only contaminants from site movements but also oils and greases from under the vehicles and, if invasive plants are present, it may help to carry them elsewhere.

## Control measures

1. Plan all activities carefully, including settlement tanks, lagoons, using grassed areas, hay bales or silt socks, and always have contingency plans in place.
2. Regular monitoring arrangements should be put in place and followed to ensure control measures are fully implemented.
3. Consider installing cut-off trenches or silt fences to prevent run-off.
4. Wash-down water must be contained and checked to ensure a pollution incident does not occur.
5. When using Bentonite ensure there are no spills of the dry powder or granules, or leakage onto the ground of the mixed material but, if it does occur, it must be cleaned up immediately.
6. Monitor weather forecasts and check regularly to ensure there are no leaks or build-ups of contaminates in the system being used.

## Precautions

1. Do not pump, over-pump or discharge without prior approval, or alter discharge arrangements without approval.
2. Do not strip land, unless it is absolutely necessary, as vegetation reduces silt run-off.
3. Do not leave Bentonite in the open air or ignore spillages.
4. Do not wash down vehicles, except in designated areas, or release water through grips.
5. Do not allow water into drains, gullies, ditches or watercourses, unless approval has been given.

When do you need a consent to discharge?
What activities on this site would generate silt, dust, or mud?
Where and how can Bentonite cause pollution?

Now inform your workforce of the company provision for water discharge from the site.
Encourage a discussion by using a real-life situation or example and ask if there are any questions.

# Fuel and oil

| | |
|---|---|
| **Reason** | Fuel spills cause damage to the environment and harm animals, plants, fish and humans. |
| **Why** | If you know how to handle fuel and keep it secure you are less likely to have a spill. |
| **Outline** | This talk covers some of the types of use and storage involved. |

*Control measures in place*

*Using an absorbent mat*

If you find a fuel or oil spill your must report it to your supervisor immediately.

# Fuel and oil

| | |
|---|---|
| **Reason** | Fuel spills cause damage to the environment and harm animals, plants, fish and humans. |
| **Why** | If you know how to handle fuel and keep it secure you are less likely to have a spill. |
| **Outline** | This talk covers some of the types of use and storage involved. |

## Use and storage

1. The most commonly found fuels and oils on site are diesel and petrol (for engines), oil (for gearboxes) and moulds, hydraulic fluid and grease.
2. Poor storage, lack of care during refuelling, vandalism and poorly maintained plant can all result in spillage.
3. Even a small spill can cause damage to the environment and cause harm to animals, plants, fish and humans, as well as contaminating watercourses and groundwater.
4. A spillage is likely to be expensive to clean up and there is the likelihood of prosecution and a large fine.

## Control measures

1. Ensure that bulk fuel and oil storage tanks are bunded with a capacity of 110% and kept secure (locked when not in use) and checked regularly.
2. All containers should be stored in secure, bunded areas.
3. Refuelling should be carried out in controlled areas, where possible, and drip trays or absorbent mats placed under static plant.
4. Spill kits should be available near the refuelling operation and drain covers should be provided.
5. Clear up minor spillages immediately and report the incident to your supervisor.
6. Seek advice before attempting to dispose of fuels, oils and contaminated water or ground materials.

## Precautions

1. Do not pour waste or wash spillages of fuel or oil down drains.
2. Do not store or carry out refuelling within 10 metres of a watercourse.
3. Do not allow drip trays to overflow or leave a tank to fill unsupervised.

What should be done when there is a spillage?
What is the maximum capacity of a bund?
How far from a watercourse should fuels and oils be stored?

Now inform your workforce of the company provision for protection against fuel and oil spills.
Encourage a discussion by using a real-life situation or example and ask if there are any questions.

# Energy and water efficiency

| | |
|---|---|
| **Reason** | Minimising waste of resources, including water, fuel and electricity, not only helps the environment but also saves money. |
| **Why** | If you know how to recognise unnecessary use and waste, you are in a better position to help reduce it. |
| **Outline** | This talk covers some of the ways of recognising the potential for reduction of wasted resources and some control measures. |

| Notes |
|---|
| |

Many tonnes of emissions are produced each year through inefficient use of resources. You can help to reduce waste.

# Energy and water efficiency

| | |
|---|---|
| **Reason** | Minimising waste of resources, including water, fuel and electricity, not only helps the environment but also saves money. |
| **Why** | If you know how to recognise unnecessary use and waste, you are in a better position to help reduce it. |
| **Outline** | This talk covers some of the ways of recognising the potential for reduction of wasted resources and some control measures. |

## Water and energy use

1. Water flows from our taps and has been through an expensive cleaning process to make it safe for us to use. Leaving a tap running or using water unnecessarily is a waste.
2. Natural resources from the ground are refined to produce products such as diesel and petrol, which we use for powering portable equipment.
3. When we plug our tools and equipment into the National Grid for electricity we are benefiting from power stations fired by everything from natural radioactive materials through to gas, coal and oil, all of which are sourced from the ground.
4. Many tonnes of emissions are produced each year through inefficient use of resources.
5. Switching off all unnecessary appliances and idling plant can make a real difference not only to the environment but to the cost of a project.

## Control measures

1. Ensure that waste is minimised by switching it off when equipment is not in use.
2. Operate plant efficiently by using the appropriate power and maintain regular services and maintenance.
3. Use locally sourced resources, equipment and materials to minimise travel.
4. Report areas where you feel improvements can be made.

## Precautions

1. Do not leave doors and windows open when the heating or air-conditioning is on – just turn it off.
2. Do not leave lights on unless they are there for security and safety reasons.
3. Do not leave plant engines running just to keep you warm in the cab.

What improvements do you think you could make to improving energy efficiency?
Would you take any of this information home and use it to reduce your fuel bills?
What steps will you take today to reduce energy consumption?

Now inform your workforce of the company provision for energy and water efficiency on site.
Encourage a discussion by using a real-life situation or example and ask if there are any questions.

# Wildlife

| | |
|---|---|
| **Reason** | Wildlife and their habitats are protected by law and must not be disturbed. |
| **Why** | Knowing which wildlife is protected and what actions to take you can avoid delays and possible prosecution. |
| **Outline** | This talk covers some of the different types of wildlife and how you can help to protect them. |

*Great crested newt*

*Lesser spotted woodpecker*

*Bats roosting*

*Badgers*

If you find wildlife near your work area you must stop, move away, and report it to your supervisor.

# Wildlife

| | |
|---|---|
| **Reason** | Wildlife and their habitats are protected by law and must not be disturbed. |
| **Why** | Knowing which wildlife is protected and what actions to take you can avoid delays and possible prosecution. |
| **Outline** | This talk covers some of the different types of wildlife and how you can help to protect them. |

## Wildlife protected by law

1. **Badgers** and their setts are protected by law. It is illegal to work near them without a licence and you must take positive steps to protect them.
2. **Bats** are an endangered species and so their habitats are protected. Some bats are only 40 mm long and can weigh as little as 5 grams. Every structure and mature tree is a potential bat roost.
3. **Great crested newts** are rare and are protected by law. They grow up to 170 mm long and can be found almost anywhere in the countryside.
4. **Adders** are the only UK poisonous snake (although a bite is rarely fatal). Snakes are cold-blooded and can often be found basking in warm locations.
5. **Nesting birds** are usually found in trees and hedgerows but can be found on the ground, in equipment or materials, eaves, and so on. The nesting season generally runs from March to the end of July, but some birds nest all year round.

## Control measures

1. Check the area before starting work and look for droppings that may indicate the presence of animals, especially if the removal of potential habitats is likely.
2. If you find wildlife present near your work area stop, move away and report it to your supervisor.
3. Try to minimise disturbance when carrying out authorised work near wildlife.
4. Follow the legal specified working distance when working near badgers and their setts.
5. Report any harm or injury to wildlife to your supervisor, who will seek professional advice.

## Precautions

1. Do not try to move, injure or harm any wildlife.
2. Do not touch any wildlife, as they can be fragile and you may cause them harm.
3. Do not move or disturb potential nesting places without first checking to ensure there is no wildlife present.

?

What time of year are you most likely to find nesting birds?
Why is it important to protect our wildlife?
What should you do if you find wildlife in a stockpile of materials?

Now inform your workforce of the company provision for wildlife protection.
Encourage a discussion by using a real-life situation or example and ask if there are any questions.

# Invasive plants

| | |
|---|---|
| **Reason** | Certain types of plant are known to be invasive and can cause harm to the environment. |
| **Why** | If you can recognise the plant and know what actions to take, this will help you prevent future problems. |
| **Outline** | This talk covers types of invasive plants and how you can help to control them. |

*Giant hogweed*

*Japanese knotweed*
*(image supplied by Bridget Plowright)*

*Himalayan balsam*

*Japanese knotweed in flower*

If you find invasive plants within seven metres of your work area stop, move away and report it to your supervisor.

# Invasive plants

| | |
|---|---|
| **Reason** | Certain types of plant are known to be invasive and can cause harm to the environment. |
| **Why** | If you can recognise the plant and know what actions to take, this will help you prevent future problems. |
| **Outline** | This talk covers types of invasive plants and how you can help to control them. |

## Types of invasive plant

1. **Giant hogweed** spreads and, by eroding the soil, can endanger the survival of native plants and be a danger to grazing animals. It contains large amounts of poisonous sap that, if exposed to sunlight, can harm humans when it comes into contact with the skin. Eye contact can cause temporary blindness.
2. **Himalayan balsam** thrives by casting off hundreds of seeds in the autumn, when the seed pods burst, that can spread to over seven metres. It dominates habitats and shades native plants from sunlight. Its dead steps block watercourses, which can cause flooding.
3. **Japanese knotweed** grows rapidly (about 20 mm per day) and the root system is likely to extend by many metres and up to a depth of two metres. It can grow through foundations, walls, roads and drainage pipework.

## Control measures

1. Check the area before starting work and look for signs of any of the three plants mentioned above, including ground or protective fencing disturbance.
2. If you find invasive plants within seven metres of your work area stop, move away and report to your supervisor.
3. Seek medical advice (from your first aider) if you come into contact with the sap of giant hogweed.
4. Follow site guidance when working near to invasive plants to prevent them spreading or harming the environment or wildlife.

## Precautions

1. Do not enter identified areas containing invasive plants.
2. Do not excavate materials that may contain seeds or material from invasive plants.
3. Do not stockpile material, especially within 10 metres of a watercourse.
4. Do not drive vehicles through invasive plants, as this will cause them to spread.
5. Do not double-handle material suspected of containing invasive plants but, if it is unavoidable, do so on an impermeable surface.

Why are invasive plants a problem in the UK?
How should giant hogweed be handled (what PPE is required)?
Why is it important not to double-handle invasive plants?
What should you do if you get giant hogweed sap on your skin or in your eye?

Now inform your workforce of the company provision for identification and prevention of spreading invasive plants.
Encourage a discussion by using a real-life situation or example and ask if there are any questions.

# Working around trees and hedgerows

| | |
|---|---|
| **Reason** | Trees and hedgerows are an important part of our environment and countryside. |
| **Why** | If you know how to protect them when working nearby you will not damage them. |
| **Outline** | This talk covers the importance of protection and the controls that need to be in place. |

Trees can become unstable and a hazard to people and property if damaged by compaction or cutting of roots, pollution and through impact by machines.

# Working around trees and hedgerows

| | |
|---|---|
| **Reason** | Trees and hedgerows are an important part of our environment and countryside. |
| **Why** | If you know how to protect them when working nearby you will not damage them. |
| **Outline** | This talk covers the importance of protection and the controls that need to be in place. |

## Trees and hedgerows

1. Trees and hedgerows provide a vitally important habitat for many different types of wildlife.
2. They are protected by legislation, which is enforced by the Local Authority.
3. At certain times in the year they may contain nesting birds, which should not be disturbed as they are protected by law.
4. Trees can become unstable and be a hazard to people and property if they are damaged by compaction or cutting of roots, pollution and through impact by machines.

## Control measures

1. Check to make sure permission has been granted before any work starts to fell trees, remove hedgerows or clear vegetation, when instructed to do so by your supervisor.
2. Excavate near trees or hedgerows with great care. Root systems must be hand dug.
3. Store fuels, oils and other potential pollutants away from root systems.
4. Follow instructions with regard to protection zones around trees.

## Precautions

1. Do not carry work out on trees or hedgerows during the nesting season.
2. Do not undertake work near to trees without clearance from your supervisor.
3. Do not track vehicles or plant over tree protection areas.
4. Do not store materials under tree canopies.

When should you not carry out work on trees or hedgerows?
What should you do if nesting birds, or birds' nests are found on site?
Where should fuels and oils be stored?

Now inform your workforce of the company provision for tree and hedgerow protection on this site.
Encourage a discussion by using a real-life situation or example and ask if there are any questions.

# Waste

| | |
|---|---|
| **Reason** | Waste management and control is a vital element in the construction industry. |
| **Why** | Every year millions of pounds are wasted by poor management of reuse and recycling techniques. |
| **Outline** | This talk covers some important aspects of waste management and control. |

**Eliminate** – Avoid producing waste in the first place.

**Reduce** – Minimise the amount of waste you produce.

**Reuse** – Use items as many times as possible.

**Recycle** – Recycle what you can only after you have reused it.

**Dispose** – Dispose of the waste to landfill.

*Dispose of waste as appropriate*

Don't skip it if you can reuse it.

# Waste

| | |
|---|---|
| **Reason** | Waste management and control is a vital element in the construction industry. |
| **Why** | Every year millions of pounds are wasted by poor management of reuse and recycling techniques. |
| **Outline** | This talk covers some important aspects of waste management and control. |

## Waste management and control

1. The golden rules are:
   **Eliminate** – Avoid producing waste in the first place.
   **Reduce** – Minimise the amount of waste you produce.
   **Reuse** – Use items as many times as possible.
   **Recycle** – Recycle what you can only after you have reused it.
   **Dispose** – Dispose of the waste to landfill.
2. Waste comes from many sources, including contaminated ground, road sweeper arisings, excavations, damaged materials, off-cuts and leftovers (mortar, concrete, plaster, plasterboard, paints, solvents, and so on). Anything not used is classed as waste.
3. Minimising waste by following the golden rules takes more effort but is good for the environment and the company.
4. Segregating waste into hazardous, non-hazardous and inert types for disposal or recycling maximises opportunities for recovery costs and penalties can be avoided.

## Control measures

1. Store materials properly and safely to prevent damage before use.
2. Keep significant off-cuts for reuse and know the correct place to stockpile and protect materials for reuse.
3. Consider the quantity of material to be used before ordering or opening a pack and use it all before opening a new pack.
4. Reuse materials (such as formwork and shuttering) where practical.
5. Tell your supervisor about instances in your work where you could reduce waste.

## Precautions

1. Do not place materials for reuse in areas where they could be damaged or be contaminated by other materials.
2. Do not use a new length of timber, pipe or cable without checking the reuseable stock.
3. Do not dispose of contaminated waste, other than in designated areas.
4. Do not overfill skips. If a skip is full tell your supervisor.
5. Do not mix hazardous, non-hazardous and inert waste together because it prevents recycling.

?

Why is it important to segregate waste?
What is the site policy for recycling and waste management?
What wastes cannot be mixed?
What happens to waste when it leaves site?
What should be done with surplus materials and off-cuts?

Now inform your workforce of the company provision for waste management and control.
Encourage a discussion by using a real-life situation or example and ask if there are any questions.

# F
# Specialist activities

# Piling

| | |
|---|---|
| **Reason** | Unless there is attention to health and safety, work on piling sites can be particularly dangerous. |
| **Why** | Piling work can involve several hazardous activities; even if not directly involved, you must take care. |
| **Outline** | This talk covers the dangers and precautions to be taken. |

*All piling operations must be controlled and supervised*

Piling is safe so long as everyone does their job and those not involved stay well clear.

# Piling

| | |
|---|---|
| **Reason** | Unless there is attention to health and safety, work on piling sites can be particularly dangerous. |
| **Why** | Piling work can involve several hazardous activities; even if not directly involved, you must take care. |
| **Outline** | This talk covers the dangers and precautions to be taken. |

## Hazards

1. Manual handling is a common feature of piling activities; many people in the industry have time off work because of resulting back injuries.
2. At times, piling activities will require operatives to work at height. Be safe if you have to do so and, even if not directly involved, be aware of what is going on above you.
3. Ejected and falling spoil can be a hazard so ensure that you wear personal protective equipment (PPE) as necessary; a safety helmet, safety boots and eye protection must always be worn.
4. Piling activities usually present fall and trip hazards from pulsing trailing hoses and low projecting pile reinforcements. The presence of these may not be obvious to the untrained.
5. Piling can be a noisy work activity. Even if not directly involved, you may have to protect your hearing and make sure you observe hearing protection signs.
6. Most piling operations will involve lifting using a crane or the piling rig itself. Keep clear if you are not directly involved.
7. Some piling operations will necessitate working near or over water. Do not interfere with safety boats or other equipment.
8. Piling can involve the use of hazardous substances. Leave them alone if you are not involved.

## Precautions

1. If involved in manual handling, get assistance if necessary.
2. Unless guard-rails are fitted, use a safety harness and lanyard, which should be clipped to a suitable secure and strong point when working at height.
3. Be aware of, and avoid, features that may cause you to slip, trip or fall; they may not be obvious at first sight.
4. Do not attempt to operate the piling rig or other equipment unless you are trained and authorised.
5. Only people who have been trained and are competent should sling loads or signal to the crane driver.
6. Never be tempted to ride on any load that is being lifted, lowered or moved.
7. If working over or near deep water, life jackets must be worn and there should be a rescue boat and trained staff available.
8. Read the Control of Substances Hazardous to Health (COSHH) assessments for any hazardous substances that you come into contact with.
9. Promptly clean off any hazardous substances that get onto exposed skin.

What should be done to prevent falls from height?
What should you do if some of your PPE is damaged?
What should you do if you discover an item of lifting equipment that you believe is damaged?

Encourage a discussion by using a real-life situation or example and ask if there are any questions.

# Road and street works

| | |
|---|---|
| **Reason** | Over 3,800 road accidents occur at road works each year. |
| **Why** | In many cases, advance warning of the works and protection of the works are inadequate. |
| **Outline** | This talk covers the precautions necessary to protect you and road users. |

*Traffic management plan*

*Operatives must wear the correct PPE*

*Signage must be clear*

# Road and street works

| | |
|---|---|
| **Reason** | Over 3,800 road accidents occur at road works each year. |
| **Why** | In many cases, advance warning of the works and protection of the works are inadequate. |
| **Outline** | This talk covers the precautions necessary to protect you and road users. |

## Consider road users

1. Ensure that appropriate warning signs are displayed and correctly positioned.
2. Cone off a tapered, lead-in zone to ease traffic movement past the works.
3. Erect barriers around all excavations and position lighting for safety at night.
4. Allow sufficient footway width for pedestrians to pass, or barrier off a temporary footway in the road.
5. Install appropriate and efficient traffic control measures.
6. Carry out work activities that create dust or debris in a manner and location so that road traffic and pedestrians are not adversely affected.
7. Position plant and equipment so that no part of it (for example, swinging jib or excavator arm) encroaches into the safety zone.
8. Do not store tools, materials, equipment or plant in the safety zone.
9. Site traffic entering or leaving the works must not endanger other road users.
10. Take precautions to prevent mud or other debris being deposited on roads or footpaths, whether within the work area or not.

## Personal safety

1. Ensure that a coned-off safety zone exists around the works.
2. Wear high-visibility clothing, safety helmet and safety footwear at all times.
3. Wear other PPE that is necessary for the job in hand as identified by your risk assessment.
4. Do not enter the safety zone in the normal course of your work.
5. Wash hazardous substances off exposed skin immediately. Barrier creams might help.
6. Protect yourself against sunburn. 95,000 cases of skin cancer occur each year.
7. Take the correct precautions when working in excavations.
8. Take the correct precautions when underground services are to be uncovered.

?

What precautions should you take when using a disc-cutter?
At what angle from the kerb should the lead-in cones be positioned?
Explain the difference between safety zone, working space and work area.
What is the most effective way of protecting against sunburn?
Where should you keep PPE when it is not being worn or used?
What action should you take if you get wet cement on exposed skin?

Encourage a discussion by using a real-life situation or example and ask if there are any questions.

# Trackside safety

| | |
|---|---|
| **Reason** | Attention to safety must be continuous and complete. |
| **Why** | The construction industry is hazardous enough; add the dangers of railway activities and the risks of injury will increase. |
| **Outline** | This talk covers some hazards and precautions associated with working on railways. |

*There are many hazards beside the track*

# Trackside safety

| | |
|---|---|
| **Reason** | Attention to safety must be continuous and complete. |
| **Why** | The construction industry is hazardous enough; add the dangers of railway activities and the risks of injury will increase. |
| **Outline** | This talk covers some hazards and precautions associated with working on railways. |

## Hazards

1. The danger of injuries resulting from the impact of moving trains cannot be overemphasised.
2. Besides direct impact, there is also the danger of being dragged under fast moving trains by the variations in air pressure as the train passes.
3. Electrified lines create extra hazards; beware electrical dangers from overhead lines and third and fourth line systems.
4. Scaffold must be properly erected to cater for the movement of trains.
5. The effects of alcohol and drugs, including some prescription drugs, will reduce your concentration and safety.
6. Deaths and severe injuries have occurred through accidental damage to buried high-voltage cables.

## Precautions

1. You must be trained and competent to work in a railway environment. Ensure you carry your track safety card or you will be refused access to the site.
2. Your Controller of Site Safety (COSS) will set up a safe system of work. Always follow the directions given, particularly with regard to safety distances and track clearance times.
3. Familiarise yourself with the differences, and restrictions imposed, when working in RED and GREEN zones.
4. Always wear the correct standard of PPE, including high-visibility clothing.
5. Always treat overhead lines and the third rail as live.
6. Remember that an electrical isolation does not guarantee safety; it does not prevent non-electric locomotives from using the track.
7. Never allow yourself or anything you are handling to come within 2.75 metres of overhead lines.
8. Exercise extreme care if handling scaffold poles or other metallic objects.
9. Ladders must be wooden or of approved non-conductive materials, never metallic.
10. If you have to remove trackside fencing, replace it as soon as is practical.
11. Do not excavate, dig or otherwise penetrate the ground surface unless a survey for underground services has shown that it is safe to do so.

?

How does the permitted amount of alcohol for trackside safety compare to that for road vehicle drivers?
When should a 'touch lookout' be used?
How long should you be in a position of safety before a train reaches you?

Encourage a discussion by using a real-life situation or example and ask if there are any questions.

# Demolition safety

| | |
|---|---|
| **Reason** | Demolition is a high risk work, with falls and premature collapse of structures as the greatest risk. |
| **Why** | Making sure you know the health and safety method statement (demolition plan) for your work can make the demolition safer. |
| **Outline** | This talk covers before and during the demolition and the hazards on a demolition site. |

*Protective measures in place*

*Damping down*

Demolition is dangerous – follow the rules.

# Demolition safety

| | |
|---|---|
| **Reason** | Demolition is a high risk work, with falls and premature collapse of structures as the greatest risk. |
| **Why** | Making sure you know the health and safety method statement (demolition plan) for your work can make the demolition safer. |
| **Outline** | This talk covers before and during the demolition and the hazards on a demolition site. |

## Before work

1. Ensure you have been briefed and know exactly what you should be doing.
2. Make sure you follow the method statement.
3. Ensure you know where demolition plant will be operating.
4. Any hazardous substances that will be disturbed must be identified and protective measures put in place.
5. Don't enter a building if it appears to be unsafe.
6. Find out from your supervisor if there are any live services.
7. Find out if there are any hazardous materials in the structure (for example, acids from industrial processes, asbestos from pipe lagging or biological hazards in old hospitals).

## During the demolition

1. Demolition must be planned, supervised and carried out by a competent person.
2. Ensure you wear your PPE.
3. Respirators are required in dusty conditions.
4. Work from correctly erected scaffold platforms or towers and hydraulic or crane-handled work baskets.
5. When cutting steel, secure gas bottles, use flashback arresters, store spare bottles in compounds, take care with hoses and provide a means of putting out fires.
6. Erect sheeting or hoardings to protect other workers or the public and replace any that are damaged.
7. Stop work and clear the area if you think you have discovered asbestos.

## Precautions

1. Don't overload floors with material, they will collapse.
2. Don't demolish walls and floors adjacent to other workers.
3. Damp down to reduce dust, and keep noise to a minimum.
4. Signallers must be provided where machines are used adjacent to other workers or the public.

What must you know before starting?
What must you do before entering a building?
What protective equipment should you be wearing?
What should you do to keep dust down?
Who should be supervising the demolition?

Now inform your workforce of the company policy regarding demolition.
Encourage a discussion by using a real-life situation or example and ask if there are any questions.

# Hydro demolition

| | |
|---|---|
| **Reason** | Hydro demolition uses a high pressure water jet to remove or cut concrete from a structure. |
| **Why** | The pressure at which the water is jetted is in itself a hazard to health and safety. |
| **Outline** | This talk covers the dangers associated with hydro demolition and how they can be overcome. |

*Beware of water at high pressure*

*You must be properly trained*

# Hydro demolition

| | |
|---|---|
| **Reason** | Hydro demolition uses a high pressure water jet to remove or cut concrete from a structure. |
| **Why** | The pressure at which the water is jetted is in itself a hazard to health and safety. |
| **Outline** | This talk covers the dangers associated with hydro demolition and how they can be overcome. |

## Hazards

1. The pressure of the water jet can be up to 2,500 bar (35,500 psi). The main danger is that of being accidentally injected by the high pressure jet.
2. The resultant injuries could be life or limb threatening. It only takes 100 psi to pierce the skin. High pressure water jet injuries are treated as surgical emergencies.
3. Tissue damage can be compared to that of a gunshot wound.
4. Being injected with contaminated water could lead to further medical complications.
5. The small entrance wound may appear minor but it must not be ignored. Report any injection injury immediately.
6. Anyone who has suffered an injection, or other injury resulting from high pressure fluids, must go to hospital immediately.
7. Equipment vibration can also cause health problems.

## What you should do

1. If someone has suffered an injection injury:
   - control any bleeding and elevate the limb where possible
   - arrange for the injured person's transfer to hospital
   - contact the hospital whilst the patient is in transit and inform them of the nature of the injury.

## Safe use

1. Never use hydro demolition equipment unless you have been properly trained.
2. Hydro demolition operations should be isolated to avoid injury to people not involved in the activity.
3. This can be done by erecting barriers and signs around the work area. In some cases, it might be possible to remove the item to be jetted to an isolated location.
4. The pressure of the water jet will try to push the person holding the nozzle backwards. A secure footing is needed.
5. If hydro demolition is taking place and you are not involved, stay well out of the area.
6. If you are involved, ensure that you have been issued with, and wear, adequate PPE.

## Equipment maintenance

1. Due to the very high pressures involved, equipment must be maintained in accordance with the manufacturer's instructions.
2. Daily pre-use checks of the equipment should be carried out by someone who has been trained and is competent to do so.
3. Do not tamper with the equipment if you do not know what you are doing; you must be properly trained.
4. Do not attempt to adjust the equipment whilst it is pressurised.

?

If during use the equipment is not operating properly, what should you do?
What are the essential items of PPE that should be worn?

Encourage a discussion by using a real-life situation or example and ask if there are any questions.

# Steel erection

| | |
|---|---|
| **Reason** | Operatives should be aware of the hazards associated with the erection of steelwork. |
| **Why** | This potentially dangerous activity poses risks to the safety of those erecting the steelwork and of others who are in the vicinity. |
| **Outline** | This talk covers the hazards associated with the erection of steelwork and the precautions required. |

*Large-scale steel erection*

*Steel erection often involves safe working at height*

# Steel erection

| | |
|---|---|
| **Reason** | Operatives should be aware of the hazards associated with the erection of steelwork. |
| **Why** | This potentially dangerous activity poses risks to the safety of those erecting the steelwork and of others who are in the vicinity. |
| **Outline** | This talk covers the hazards associated with the erection of steelwork and the precautions required. |

## Hazards

1. Falls from height are a common source of serious and fatal injuries in the construction industry. Ensure that you are never at risk of falling.
2. Falling tools and materials are a hazard to others when you are working at height.
3. Electrocution from live overhead electrical cables may be a hazard.
4. There is always a danger of impact injuries, including head injuries, when beams are being lifted.
5. If mobile elevating work platforms (MEWPs) or other access equipment is to be used, the ground conditions must be suitable.
6. Operators and others in the basket of a cherry picker must wear a harness and restraint lanyard clipped to the machine's secure anchorage point.

## Precautions

1. Erection will invariably involve the use of a crane. All lifts must be supervised by a competent person and involve the use of qualified slingers.
2. When working at height, work from a stable working platform wherever possible.
3. When a platform is not practical, wear a safety harness and fall arrest device. Ensure that you are clipped to a secure anchorage point at all times.
4. Ensure that there is a safe means of access to high level places of work.
5. Be aware of the dangers to others below and cordon off the area at ground level.
6. The use of cranes over long periods will mean that you may have to consider:
   - whether the ground conditions can support the crane
   - the area required by the crane as it slews, including, in some cases, consideration for the general public
   - the proximity of buried ducts or pipes that may affect crane stability.
7. Always wear the appropriate PPE.
8. Don't move along beams by straddling unless absolutely necessary and you are clipped to a proper anchorage point – this is a last resort option.

?

What measures should be taken to avoid contact with overhead cables?
How can the length of time spent working at height be reduced?
What should you consider as your next choice of working at height when it is not practical to erect scaffold?
What type of safety harness should be worn, and why?

Encourage a discussion by using a real-life situation or example and ask if there are any questions.

# Water jetting

| | |
|---|---|
| **Reason** | Water jetting can be a dangerous activity, if it is not carried out in a controlled manner. |
| **Why** | High pressure water, the equipment and additives used can cause serious injury. |
| **Outline** | This talk covers the dangers associated with water jetting and how they can be overcome. |

*Working alone is hazardous*

*Ensure appropriate signage and barriers are in place*

# Water jetting

| | |
|---|---|
| **Reason** | Water jetting can be a dangerous activity, if it is not carried out in a controlled manner. |
| **Why** | High pressure water, the equipment and additives used can cause serious injury. |
| **Outline** | This talk covers the dangers associated with water jetting and how they can be overcome. |

## Hazards

1. The water jet is usually at a pressure above 140 bar (over 2,000 psi). A misdirected jet hitting a person would cause serious injury or death.
2. Flying debris can also be a cause of injury.
3. Some chemical additives used are hazardous to health.
4. Equipment vibration can also result in health problems.

## Safe use

1. Never use water jetting equipment unless you have been properly trained, preferably as part of a team.
2. Most water jetting is carried out by teams of three operatives. If you are managing with fewer people is the work method safe?
3. Water jetting operations should be isolated to avoid injury to people not involved in the activity.
4. Isolation can be achieved by erecting barriers and signs around the work area, or preferably by removing the item to be jetted to an isolated location.
5. The pressure of the water jet will try to push the person holding the nozzle backwards. A secure footing is needed.
6. If working at height on a scaffold platform, ensure that you are safe. It may be necessary to wear a safety harness and lanyard, which is clipped to a secure point.
7. Ensure that you have been issued with, and wear, adequate PPE.
8. Read the COSHH assessments for any substances used or disturbed.

## Equipment maintenance

1. Due to the high pressures involved, equipment must be maintained in accordance with the manufacturer's instructions.
2. If any part of the equipment is assembled incorrectly, high pressure leaks could cause serious personal injury.
3. Daily pre-use equipment checks must be carried out by someone who has been trained and is competent to do so.
4. Do not tamper with the equipment if you do not know what you are doing; you must be properly trained.
5. Do not attempt to adjust the equipment whilst it is pressurised.

?

If the item to be jetted cannot be moved to a remote location, what other measures might be used to keep people at a safe distance?
Why should you not use water jetting on asbestos?
If working at height, why is it particularly important that scaffold working platforms are completely stable?
What essential items of PPE should you wear?

Encourage a discussion by using a real-life situation or example and ask if there are any questions.

# Further information

# Toolbox talk delivery log

| A Legal and management | | | | |
|---|---|---|---|---|
| **Toolbox talk** | **Date** | **Date** | **Date** | **Date** |
| A01 | | | | |
| A02 | | | | |
| A03 | | | | |
| A04 | | | | |
| A05 | | | | |
| A06 | | | | |
| A07 | | | | |
| A08 | | | | |
| A09 | | | | |
| A10 | | | | |
| A11 | | | | |
| A12 | | | | |
| A13 | | | | |

| B Health and welfare | | | | |
|---|---|---|---|---|
| **Toolbox talk** | **Date** | **Date** | **Date** | **Date** |
| B01 | | | | |
| B02 | | | | |
| B03 | | | | |
| B04 | | | | |
| B05 | | | | |
| B06 | | | | |
| B07 | | | | |
| B08 | | | | |
| B09 | | | | |
| B10 | | | | |
| B11 | | | | |
| B12 | | | | |
| B13 | | | | |
| B14 | | | | |
| B15 | | | | |
| B16 | | | | |
| B17 | | | | |
| B18 | | | | |
| B19 | | | | |

## C General safety

| Toolbox talk | Date | Date | Date | Date |
|---|---|---|---|---|
| C01 | | | | |
| C02 | | | | |
| C03 | | | | |
| C04 | | | | |
| C05 | | | | |
| C06 | | | | |
| C07 | | | | |
| C08 | | | | |
| C09 | | | | |
| C10 | | | | |
| C11 | | | | |
| C12 | | | | |
| C13 | | | | |
| C14 | | | | |
| C15 | | | | |
| C16 | | | | |
| C17 | | | | |
| C18 | | | | |
| C19 | | | | |
| C20 | | | | |
| C21 | | | | |

## D High risk activities

| Toolbox talk | Date | Date | Date | Date |
|---|---|---|---|---|
| D01 | | | | |
| D02 | | | | |
| D03 | | | | |
| D04 | | | | |
| D05 | | | | |
| D06 | | | | |
| D07 | | | | |
| D08 | | | | |
| D09 | | | | |
| D10 | | | | |
| D11 | | | | |
| D12 | | | | |

## E Environment

| Toolbox talk | Date | Date | Date | Date |
|---|---|---|---|---|
| E01 | | | | |
| E02 | | | | |
| E03 | | | | |
| E04 | | | | |
| E05 | | | | |
| E06 | | | | |
| E07 | | | | |
| E08 | | | | |
| E09 | | | | |
| E10 | | | | |
| E11 | | | | |

## F Specialist activities

| Toolbox talk | Date | Date | Date | Date |
|---|---|---|---|---|
| F01 | | | | |
| F02 | | | | |
| F03 | | | | |
| F04 | | | | |
| F05 | | | | |
| F06 | | | | |
| F07 | | | | |

# Notes

| Toolbox talk reference | Notes, comments or supporting information |
|---|---|
| | |
| | |
| | |
| | |
| | |
| | |
| | |
| | |
| | |
| | |
| | |
| | |
| | |
| | |
| | |
| | |
| | |
| | |
| | |
| | |
| | |
| | |
| | |
| | |
| | |
| | |
| | |
| | |
| | |
| | |
| | |

# Feedback form

| | |
|---|---|
| Toolbox talk topic | |
| Feedback and points learnt from delivering the toolbox talk | |
| Ways that I can improve my approach generally or on the delivery of this topic | |

| | |
|---|---|
| Toolbox talk topic | |
| Feedback and points learnt from delivering the toolbox talk | |
| Ways that I can improve my approach generally or on the delivery of this topic | |

| | |
|---|---|
| Toolbox talk topic | |
| Feedback and points learnt from delivering the toolbox talk | |
| Ways that I can improve my approach generally or on the delivery of this topic | |

# Briefing record

| | | | |
|---|---|---|---|
| **Project/site** | | | |
| **Toolbox talk topic** | | | |
| **Date** | | **Time** | |
| **Briefing delivered by** | | **Signature** | |
| **Company** | | | |

List of persons attending briefing (obtain signatures at end of briefing as a record of attendance).

| Name | Company | Signature |
|---|---|---|
| | | |
| | | |
| | | |
| | | |
| | | |
| | | |
| | | |
| | | |
| | | |
| | | |
| | | |
| | | |
| | | |

| Workforce feedback and comments |
|---|
| |